climate change and insurance

DISASTER RISK FINANCING IN DEVELOPING COUNTRIES

GUEST EDITOR:
Eugene N. Gurenko

climate policy

VOLUME 6 ISSUE 6 2006

earthscan
from Routledge

First published by Earthscan in the UK and USA in 2007

For a full list of publications please contact:
Earthscan
2 Park Square, Milton Park, Abingdon, Oxfordshire OX14 4RN
Simultaneously published in the USA and Canada by Earthscan
52 Vanderbilt Avenue, New York, NY 10017

First issued in paperback 2018

Earthscan is an imprint of the Taylor & Francis Group, an informa business

ISSN: 1469-3062
ISBN-13: 978-1-84407-483-9 (hbk)
ISBN-13: 978-1-138-38410-1 (pbk)

Typeset by Domex

Cover design by Paul Cooper Design

Climate Policy is the leading international peer-reviewed journal on responses to climate change. For further information see www.climatepolicy.com.

Contents

List of contributors

Christoph Bals is the Executive Director and the founding member of Germanwatch – a non-profit organization founded in 1991. He was among the initiators of the European Business Council for Sustainable Energy, the pro-Kyoto-campaign 'e-mission55', and the Initiative for Climate Conscious Flying Atmosphere. He has been a Board member of the Foundation for Sustainability since 1998. He has been one of the three NGO representatives in the German government's Working Group on Emission Trading (AGE) since 1998; and a member of the advisory group of the German Green Investment Index (NAI); in 2003 and 2004 he was on the National Advisory Committee for Renewables 2004. Christoph has headed several different successful environmental campaigns (Rio Konkret, Climate Responsibility Campaign). He studied theology, economics and philosophy at Munich, Belfast, Erfurt and Bamberg.

Joanne Linnerooth-Bayer is leader of the Risk and Vulnerability Programme at the International Institute of Applied Systems Analysis (IIASA) in Laxenburg, Austria. She is an economist by training, and has received a BS and PhD at Carnegie-Mellon University and the University of Maryland, respectively. Her current interest is global change and the vulnerability of developing countries to catastrophic events, and she is investigating options for improving the financial management of catastrophe risks on the part of households, farmers and governments in transition and developing countries. She has recently led research projects on this topic in the Tisza River region, Hungary, and the Dongting Lake region, China. Joanne is currently a leader of two work programmes on an integrated European Union project, which examines risk and vulnerability to weather-related extremes in Europe. She is an associate editor of the *Journal for Risk Research* and on the editorial board of *Risk Analysis* and *Risk Abstracts*. She has received the Distinguished Scientist Award from the European Society for Risk Analysis and the Scientific Excellence Award from the International Society for Risk Analysis.

Sonja Butzengeiger is an expert in Kyoto Mechanisms and EU Emissions Trading. She has been working on climate policy aspects since 1999. From 2000 to 2003 she worked on a research project on baseline standardization and accounting issues (PROBASE) for the EU. Besides the implementation of CDM and JI into business practice, her focus is company-level emissions trading schemes such as the EU-ETS. Since 2001, Sonja has been working for the German Emissions Trading Group (AGE) under the lead of the German Ministry for the Environment. She also has extensive experience with strategies for the allocation process from the business perspective, the establishment of CO_2-monitoring plans, and the identification of internal GHG reduction potential by innovative approaches. She holds an engineering degree in environmental sciences.

Andrew Dlugolecki is now a Visiting Research Fellow at the Climatic Research Unit, University of East Anglia, an Advisory Board member at the Tyndall Centre for Climate Change Research and the Carbon Disclosure Project, and advisor on climate change to the UNEP Finance Initiative. He worked for 27 years in the insurance group Aviva in various senior technical and operational posts, retiring from the position of Director of General Insurance Development in December 2000. He served as the chief author and later reviewer for the Intergovernmental Panel on Climate Change in its Second, Third and Fourth (due 2007) Assessment Reports, carried out similar duties for official UK and EU reviews of climate change, and chaired two studies of climate change by the Chartered Insurance Institute (1994 and 2001). He is the author of numerous international publications on insurance and climate change. Andrew obtained a BSc (Hons) in pure mathematics at Edinburgh (1970), an MA in operational research at Lancaster (1971), and a PhD in technological economics at Stirling (1978), and is a Fellow of the Chartered Insurance Institute (1990), and a Fellow of the Royal Meteorological Society (1992).

Eugene N. Gurenko is a Lead Insurance Specialist at the Insurance Practice of the World Bank. During his career at the World Bank Group, which he joined in 1998, he designed and managed the World Bank programme of technical assistance and lending to the Turkish Catastrophe Insurance Pool (TCIP), currently one of the largest and most successful public–private partnerships in earthquake insurance in the world. His most recent projects include risk assessments and the design of catastrophe risk transfer solutions for Romania, Bulgaria, Iran, the Caribbean islands and India. In 2004–2006, Eugene was with Munich Re on a special staff exchange assignment, where he led the company's efforts to develop a group-wide terrorism risk management strategy. He is the author of numerous publications on catastrophe insurance. Eugene holds a PhD from Columbia University and professional designations of Chartered Property Casualty Underwriter (CPCU) and Associate in Reinsurance (ARe).

Erik Hoekstra is a member of and deputy to the head of the Innovative Solutions Team in Munich Re's Global Clients Division. After studying at the University of Groningen in the Netherlands, he held various positions within the Gerling Group in Cologne and Luxembourg and was an underwriter at Converium in Cologne. Before joining Munich Re in October 2002, Erik was an associate director in the Risk Markets Group at WestLB in Düsseldorf. He holds a MS degree in economics.

Peter Hoeppe is the Head of the Geo Risks Research and Environmental Management Department at Munich Re. He joined Munich Re in 2004 after a long and successful academic career. During his time at the university, he received PhD degrees in physics (1984) and human biology (1996) and was appointed Professor in 2003. Since 1984, he has held the positions of tenured Lecturer at the Institute of Bioclimatology and Applied Meteorology and the Institute of Occupational and Environmental Medicine at the Ludwig-Maximilians-University (LMU) in Munich, Germany. His main research fields have been health effects of weather and climate, as well as pollutants and the general assessment of environmental risks. On a couple of occasions, Peter has worked abroad in the USA, Austria, and Pakistan. He is a member of the International Society of Biometeorology, of which he was President in 1999–2002, and a member of the German Meteorological Society, where for many years he served on the Board.

Catherine Rose James has nearly 2 years of research/consulting experience in rural development projects, particularly those relating to the water and sanitation sector. Her key focus areas have been policy and institutional analysis and impact assessment of water resources projects, and her area of expertise includes social mobilization using participatory tools, microfinance and data analysis. Prior to being associated with TERI, she worked as a risk underwriter with State Bank of India Cards, where she undertook risk analysis and underwriting of SBI classic international credit cards and portfolio health assessments. Catherine is a graduate in economics from St Stephens College, Delhi, and holds a post-graduate diploma in rural management from Xavier Institute of Management, Bhubneshwar (2004).

Ulka Kelkar is an Associate Fellow with the Centre for Global Environment Research, The Energy and Resources Institute (TERI), India. She has a Masters degree in economics from the School of International Studies, Jawaharlal Nehru University, New Delhi. She has more than 6 years work experience in the fields of climate change negotiations and policy, clean development mechanism projects, and vulnerability and adaptation assessment. Her recent projects include a review of the preparedness of the Indian insurance industry to climate change, research on key negotiating issues for India, a national strategy study on CDM for India, and analysis of the role of emissions trading in climate policy.

Ritu Kumar is an environmental economist experienced in dealing with sustainable development issues, energy and climate change. She is Director of The Energy and Resources Institute (TERI) office in London. She is currently working on a number of practical projects aimed at enhancing the capacity of developing-country partners to participate in potential Kyoto Protocol mechanisms. One of these projects is looking at the future role of the Indian insurance industry in climate change. She has worked with the United Nations Industrial Development Organization (UNIDO) for 10 years, of which 2 years were spent in West Africa, and has developed and implemented projects relating to industry and environmental policy in several developing countries. Ritu is a postgraduate in economics from the Delhi School of Economics, India, and the London School of Economics.

Reinhard Mechler is a Research Scholar in the Risk and Vulnerability Programme at IIASA. He has been analysing the impacts and costs of natural disasters in developing countries, as well as strategies to reduce these costs; in particular, strategies related to risk financing. He has also studied the costs, impacts and benefits of reducing the effects of air pollution and climate change. He has worked as a consultant for the ProVention Consortium, the World Bank, the Inter-American Development Bank, and the Gesellschaft für Technische Zusammenarbeit (GTZ). Reinhard studied economics, mathematics and English. He holds a diploma in economics from the University of Heidelberg and a PhD in economics from the University of Karlsruhe in Germany.

Axel Michaelowa is the Head of the Research Group on International Climate Policy at Zurich University and has a 12-year background in the analysis of climate policy instruments. From 1999 to 2006 he headed the climate policy programme of the Hamburg Institute of International Economics. He is also CEO of the consultancy Perspectives Climate Change, which specializes in CDM and emissions trading. He is a member of the CDM Executive Board's Registration and Issuance Team and on the UNFCCC roster of experts on baseline methodologies, where he has reviewed ten proposed methodologies. Axel has written over 50 publications on the Kyoto Mechanisms, including a book on CDM's contribution to development. He is a lead author in the Fourth Assessment Report of the Intergovernmental Panel on Climate Change and a member of the board of the Swiss Climate Cent Foundation.

Koko Warner is a senior scientific advisor at the United Nations University Institute for Environment and Human Security (UNU-EHS). She coordinates the Munich Re Foundation Chair on Social Vulnerability. Prior to joining the UNU, she was an economic research scholar at the Natural Hazards Department at the World Institute for Disaster Risk Management (DRM) in Davos, Switzerland. Koko has worked for the past 7 years on the economic and societal impacts of climate change and natural catastrophes in developing countries, with the major emphasis on the development of policy and financial instruments to reduce and transfer disaster risk. Koko received her doctoral degree in economics, and currently also serves as an Assistant Professor on the University of Richmond's Emergency Service Management graduate programme.

Foreword

Peter Hoeppe

Head of Geo Risks Research, Munich Re, Munich, Germany

Over the last few decades, both the frequency of large natural disasters as well as the amount of damage caused by them have increased significantly. 2005 was not only the second warmest year since 1856 but also a year of absolute records in number and intensity of hurricanes in the North Atlantic as well as in the global economic and insured losses caused by weather-related disasters. In recent years, science has provided more and more evidence that there is a high probability of a causal correlation between climate change and these trends in natural catastrophes. If the scientific global climate models are accurate, the present problems will be magnified in the near future. Changes in many atmospheric processes will profoundly impact upon the lives, health and property of millions of people.

The crucial question today is not when we will have the ultimate proof for anthropogenic climate change, but which strategies we should follow to mitigate and adapt to climate change. Insurance-related mechanisms can be an effective part of adaptation strategies. In particular, developing countries are very vulnerable to these changes as in these countries natural catastrophes can cost a large proportion of their GDP and consume large amounts of the money donated by developed countries that is then not available for investments in economic development.

In response to the growing realization that insurance solutions can play a role in adaptation to climate change, as suggested in paragraph 4.8 of the Framework Convention and Article 3.14 of the Kyoto Protocol, the Munich Climate Insurance Initiative (MCII) was founded in April 2005. The members of this initiative are representatives of the insurance and reinsurance industry, climate change and adaptation experts, NGOs, and policy researchers. MCII introduced and discussed its objectives for the first time in public at a special side-event of the COP-11 conference in Montreal in December 2005. This special issue of *Climate Policy* draws, by and large, on the results of the first year's work of MCII. The publication of these articles is intended to stimulate discussion on insurance-related mechanisms and how they can help in adapting to a changing climate and the corresponding risks.

Introduction and executive summary

Eugene N. Gurenko*

World Bank, Washington DC, US

1. Objectives of the publication

The increasing frequency and severity of extreme weather events (including heatwaves, droughts, bush fires, tropical and extratropical cyclones, tornadoes, hailstorms, floods and storm surges) and the historically unprecedented economic losses observed in 2004/5 have intensified the ongoing international debate about the possible adverse impact of climate change on global weather patterns. However, the adverse implications of climate change are likely to vary considerably from one country to another based on geographical location, effectiveness of climate adaptation strategies, level of insurance penetration, and the overall resilience of the economy to exogenous shocks. While the complexity of these atmospheric phenomena makes it difficult to accurately predict the impact of climate change on a given country, it is clear that disaster-prone developing countries are likely to be affected most severely due to their weaker economic base and the very limited use of risk transfer instruments in these societies.

Catastrophe risk transfer from disaster-prone countries to global reinsurance and capital markets represents one viable adaptation solution which has been gaining the support of international financial organizations. Article 4.8 of the United Nations Framework Convention on Climate Change (UNFCCC) and the supporting Article 3.14 of the Kyoto Protocol call upon developed countries to consider actions, including insurance, to meet the specific needs and concerns of developing countries in adapting to climate change. However, to date, there has been little understanding or agreement within the climate change community on the role that insurance-based mechanisms can play in assisting developing countries to adapt to climate change.

Responding to this low level of awareness of the role that can be played by insurance-related mechanisms in countries' climate change adaptation strategies, a group of NGOs, reinsurers, climate-change and insurance experts from international financial organizations, and policy researchers from academic think-tanks decided to form the Munich Climate Insurance Initiative (MCII). Founded in 2005, the organization provides an open forum for examining insurance-related options that can assist with adaptation to the risks posed by climate change. Among the most well known organizations that comprise the MCII membership are the World Bank, the United Nations, Munich Re, Germanwatch, IISA, the Potsdam Institute for Climate Impact Research (PIK) and the Swiss Federal Institute of Technology (SLF).

* Tel.: +1-202-458-5414; fax: +1-202-614-0920
E-mail address: egurenko@worldbank.org

This special issue of *Climate Policy* is the first collective publication by MCII members. It presents articles on the topic of insurance and climate change in developing countries. The issue aims to help communities at risk, governments, international organizations, the insurance industry and NGOs worldwide that are seeking solutions for preventing and adapting to the increasingly adverse economic impacts of climate change and weather-related disasters in developing countries.

The publication pursues two main objectives. First, it aims to shed light on the rationale and potential applications of catastrophe risk transfer mechanisms (insurance) for mitigating the adverse economic consequences of climate change on disaster-prone developing countries. Second, it attempts to engender an international debate on the role of insurance-based mechanisms in reducing global emissions and encouraging climate-friendly corporate behaviour.

The structure of the special issue is as follows. Hoeppe and Gurenko first discuss the scientific and economic rationale for a climate change insurance-based adaptation system. They examine the role of insurance in reducing the long-term vulnerability and mitigating the adverse financial effects of climate change on the economies of disaster-prone developing countries. They also describe the current global model of disaster risk financing and highlight its major drawbacks. Linnerooth-Bayer and Mechler provide a detailed overview of the existing public–private partnerships in catastrophe insurance and lay out an alternative design for a global climate risk financing vehicle. Bals, Warner and Butzengeiger introduce yet another alternative approach to the design of the climate change financing mechanism and discuss how it can be financed. Dlugolecki and Hoekstra present the perspective of the private sector on public–private partnerships in catastrophe risk management and describe how the competencies and resources of the global reinsurance industry can be best employed in support of such an undertaking. Kelkar, James and Kumar present a case study of traditional and innovative climate risk financing products in India, with extensive comments on their affordability and effectiveness. Michaelowa assesses the feasibility of applying insurance solutions to mitigate the negative impacts of global adaptation policies on the economies of oil exporting countries. The final article concludes and offers specific policy recommendations on how insurance-based mechanisms can be used to meet the needs and concerns of countries in adapting to climate change.

2. Executive summary

Peter Hoeppe and Eugene Gurenko offer the scientific and economic rationales for innovative climate insurance solutions in the context of global adaptation to climate change. The arguments presented in their article are twofold. On the one hand, drawing on the growing body of scientific evidence that climate change is already taking place, the authors point out that the increasing frequency and intensity of weather-related hazards makes the previous disaster-funding approaches obsolete. Indeed, according to the World Meteorological Organization (WMO), the last 5 years (2001–2005) were among the six warmest recorded worldwide since 1861, with 2005 being the second warmest. The year 2005 also set records for hurricanes in the North Atlantic: since records have been kept (1850) there have never been so many named tropical storms developing so early in the season (seven by the end of July), and the total number of 27 easily outstrips the old record of 21. Hurricane Wilma achieved the lowest recorded central pressure, and Hurricane Katrina was the most expensive ever. Already today, increasing losses from natural disasters make it more and more difficult for disaster-prone nations to finance economic recovery from their own budget revenues or special government disaster funds. All these manifestations of increasing climate extremes make a good case for insurance-based climate risk financing mechanisms at the country level. For a fixed premium payment, countries can cap the amount of fiscal loss caused by natural disasters in the future. Hence, by adopting insurance-based funding solutions, countries can not only greatly reduce the uncertainty of national budgetary outcomes due to natural disasters but can also increase the speed of post-disaster economic recovery.

The authors point out that, due to limited tax bases, high indebtedness and low uptake of insurance, many highly exposed developing countries cannot fully recover from slow- and sudden-onset disasters by simply relying on external donor aid, which typically covers only a small fraction of total economic loss. A concern to donors and multilateral financial institutions, among others, is that the increasing share of aid spent on emergency relief and reconstruction stifles spending on social, health and infrastructure investments and distorts countries' incentives for engaging in *ex-ante* risk management. This means that as disasters continue to profoundly impact on the lives, health and property of millions of people, their devastating impacts will be felt most by the world's poor. To date, these vulnerable groups have also had the least access to affordable insurance. In the absence of new innovative global disaster risk financing mechanisms, which can address the increasing volatility and severity of losses sustained by these economies due to natural disasters, and which, at the same time, can provide appropriate incentives for *ex-ante* risk management for disaster-prone countries and their populations, the adverse impact of the global climate change is likely to become even more pronounced in the future.

Joanne Linnerooth-Bayer and Reinhard Mechler lay out their vision for an international public–private climate risk insurance fund. They suggest a two-tiered climate insurance strategy that would support developing country adaptation to the risks of climate variability and meet the intent of Article 4.8 of the United Nations Framework Convention on Climate Change (UNFCCC). The core of this strategy is the establishment of a climate insurance programme specializing in supporting developing country insurance-related initiatives for sudden- and slow-onset weather-related disasters. This programme could take many institutional forms, including an independent facility, a facility in partnership with other institutions of the donor community, or as part of a multi-purpose disaster management facility operated outside of the climate regime. Its main purpose would be to enable the establishment of public–private safety nets for climate-related shocks by assisting the development of (sometimes novel) insurance-related instruments that are affordable to the poor and coupled with actions and incentives for proactive preventative measures. A second tier could provide disaster relief contingent on countries making credible efforts to manage their risks. Since it would be based on precedents of donor-supported insurance systems in developing countries, one main advantage of this proposed climate insurance strategy is its demonstrated feasibility. Other advantages include its potential for linking with related donor initiatives, providing incentives for loss reduction, and targeting the most vulnerable. Although many details and issues are left unresolved, it is hoped that this suggested strategy will facilitate much-needed discussions on practical options for supporting adaptation to climate change in developing countries.

In their contribution, the authors draw extensively on their international experience in public–private partnerships in catastrophe risk transfer, which they use to illustrate the types of country-based risk financing programmes such as those that an international facility can support.

Christoph Bals, Koko Warner and Sonja Butzengeiger provide yet another interesting proposal for insuring the uninsurable. The proposed design features of the Climate Change Financing Mechanism (CCFM) aim to rectify numerous deficiencies of the existing model of disaster aid. One of the key problems with the current *ex-post* and *ad-hoc* form of international assistance is that it neither requires nor provides any incentives for effective risk reduction or climate adaptation measures on the part of aid-receiving countries. In the absence of effective risk reduction/adaptation measures, the increasing frequency and severity of natural disasters due to climate change is likely to claim even higher future tolls in terms of economic damage and lives lost in disaster-prone developing countries.

The authors propose the establishment of a clearly defined contractual arrangement between the insurance fund and the insured countries. The fund would provide catastrophe insurance cover to countries that are highly exposed to the risk of natural disasters on a parametric basis, although free of charge in order to make such coverage affordable. Instead of paying a monetary premium, countries would be required to make an in-kind contribution commensurate with the level of their imputed risk-based premium by investing in risk reduction and mitigation projects that over time will reduce their vulnerability to future natural disasters.

The extent of adaptation measures needed to qualify a country for the CCFM basic cover would depend on its risk profile as well as its financial capacity. By encouraging more risk-prone countries to invest relatively more in risk-reduction projects, the CCFM mechanism would be providing strong incentives for proactive risk reduction. Climate adaptation measures through investments in emission-reducing projects and technology would also count toward the country's in-kind premium contributions to the CCFM.

While the main objective of the CCFM mechanism is to provide coverage for the most extreme catastrophic natural events, it may also offer an additional insurance coverage that would cover damages below the level of attachment of the basic free-of-charge insurance coverage. Such extra coverage would be provided for an additional risk-based premium to be paid by countries directly to the fund.

Similarly to the IISA proposal, financial support from the international community would be required to either subsidize the countries' risk reduction projects and/or to provide risk capital for CCFM to reduce its costs of reinsurance and consequently the costs of CCFM's coverage. Among the possible sources of CCFM's financing, the authors see financial contributions by UNFCCC Parties and by international financial organizations committed to developing sustainable climate adaptation mechanisms in disaster-prone countries.

In conclusion, the authors emphasize that the proposal can be used by the UNFCCC Parties as the base for developing the legal and organizational framework for the post-2012 Kyoto Protocol negotiations.

Andrew Dlugolecki and Erik Hoekstra offer an insurance industry perspective on the role of the private sector in insuring climate-related hazards in the context of climate change. The authors begin with an overall discussion of the role of the private sector and the key actors in the global catastrophe risk market. The complexity of the insurance market necessitates the presence of many different players which, as well as insurers and reinsurers, includes brokers, risk modellers, loss adjustors, customer associations, banks and, more recently, investors.

Although many national insurance markets and the global reinsurance and capital markets are already active in providing cover against natural catastrophes, the overall insurance market appetite for catastrophic risk is limited by companies' internal risk management considerations. Hence, commercial insurers are reluctant to provide cover for floods, windstorms and other potentially high-consequence climate events, if it involves risks with a considerable loss accumulation potential and for which hardly any historical data exist.

However, the main stumbling block to the expansion of catastrophe insurance coverage offered by the private markets is that often catastrophe insurance cover is not affordable or accessible to poor nations or individuals. This problem, however, can potentially be addressed by the creation of public–private partnerships (PPP) or through donor support for insurance-based risk financing mechanisms.

The authors then examine the type of arrangements that would provide the best fit for both public and private sector participation in catastrophe risk insurance. Their article briefly reviews a range of core and support functions essential for the successful operation of a catastrophe insurance entity before zooming in on the main competencies of the public and private sectors.

The authors point out that among the key public functions in catastrophe risk management are effective risk prevention and risk reduction, which can be achieved by the vigorous enforcement of construction codes and hazard-linked land zoning, based on thorough public risk assessment surveys. A breakdown in the implementation of these essential hazard risk management functions by governments creates additional uncertainty for private risk underwriters and results in higher risk premiums for insurance coverage.

Potentially, in a PPP, the private sector can fulfil some risk-bearing and many essential non-risk-bearing functions. In the case of the risk-bearing function, PPPs may find it advantageous from the risk management perspective to cede at least a part of their catastrophe risk peak accumulations to the global reinsurance or capital markets. Examples of such risk transfers from public–private insurance entities to the reinsurance markets are readily available around the globe. The non-risk-bearing functions of the private sector may include technical support for risk assessment, risk management, product design, distribution, marketing, loss handling and administration. A fruitful approach to explore is a PPP where the public sector sets a rigorous framework to control and reduce the physical risks, and also provides cover for severe but unlikely catastrophe events or for segments of the market which require high administration costs (due to the lack of the existing private insurance infrastructure, for example), while the private sector provides insurance services and coverage for less severe but more frequent events to the segments of economy that are more easily accessible.

The article then briefly comments on the feasibility of different PPP design approaches, including the type of insurance coverage to be provided by such entities and the level of risk aggregation (global versus regional versus local) at which they may operate. Having assessed potential design options for PPPs in catastrophe insurance, the authors conclude that the fundamental building block is the national (country) level, since risks must be consistently estimated and dealt with in their everyday context prior to their aggregation at supranational level within regional or global markets.

Ulka Kelkar, Catherine Rose James and Ritu Kumar present a case study of India's insurance industry in the context of climate change, which is typical of most other poor countries. The authors demonstrate that, given the country's history of disaster losses compounded by the growth in population concentrations and the burgeoning development in coastal and flood-prone areas, the potential impact of climate change on the Indian economy can be quite severe. These findings are driven home by the July 2005 floods in Mumbai, India's commercial capital, caused by a record level of 944 mm precipitation within 24 hours. The floods resulted in the record economic loss of US $5 billion and 1,130 people killed.

Yet, despite being the second most disaster-prone country in the world, India remains a country where insurance penetration for natural hazards is almost non-existent, less than 1%, which is abysmally low even when compared with countries with a similar level of GDP. In India, partly as a result of such a low level of insurance coverage, the government by and large remains the main financier of disaster relief, rescue, rehabilitation and reconstruction efforts.

The low insurance penetration in the country can be traced to a number of factors. On the demand side, the biggest hurdles are the lack of insurance awareness among the public and the very low income of the population. As a result, personal risk management is usually reactive and, in the case of natural catastrophes, episodic. The experience of major insurance companies shows that following a major catastrophe, first there is a rush to buy insurance cover, but this interest is short-lived and in most cases these policies are not renewed.

The scalability of successful insurance projects is further limited by the lack of incentives to purchase insurance on the part of consumers, as the government and other donor agencies often compensate losses on account of disasters. Such government assistance, however, is often insufficient or comes too late to make a real difference for the poor. As a result, as traditional risk-sharing strategies break down in the case of natural disasters that affect whole communities at once, the rural poor are forced to turn to moneylenders or sell their productive assets, which frequently undermines the very prospect of recovering their livelihoods.

Traditionally, due to the very limited insurance penetration, the insurance industry in India has played a very marginal role in dealing with the impacts of either climate variability or extreme events such as droughts, floods and cyclones. However, the recent partial liberalization of the Indian insurance market has opened the door for product innovation. Various innovative products, including those aimed at dealing with the risk of climate variability, have been introduced. Among these new products are index-based weather risk insurance contracts, which have emerged as a promising alternative to traditional crop insurance. These are linked to the underlying weather risk defined by an index based on historical data (e.g. for rainfall, temperature, snow, etc) rather than the extent of loss (e.g. crop yield loss). As the index is objectively measured and is the same for all farmers, the problem of moral hazard is minimized, the need to draw up and monitor individual contracts is avoided, and the administration costs are reduced. Weather-indexed insurance can help farmers avoid major downfalls in their overall income due to adverse weather-related events. This improves their risk profile and enhances access to bank credit, and hence reduces their overall vulnerability to climate variability. Unlike traditional crop insurance, where claim settlement may take up to a year, quick payouts in private weather insurance contracts can improve recovery times and thus enhance the farmers' coping capacity.

However, one of the main inherent disadvantages of weather derivatives is that, because of the way the index is defined, there may be a mismatch between payoffs and the actual farmer's losses; the problem also known as a basis risk. Despite many technical advantages of index-based weather risk derivatives, the presence of the basis risk makes buyers vulnerable to the possibility of not receiving compensation in spite of suffering a considerable loss, which makes these instruments ill-suited for small farmers. The problem of the basis risk, however, becomes less pronounced for commercial buyers of these instruments (such as large commercial farmers, agricultural lenders and farmers' cooperatives) due to the diversification effect afforded by their larger land-holdings and their higher risk retention capabilities.

The authors conclude that in achieving this goal the private insurance industry would benefit from joining forces with the government in the form of a PPP. Such an alliance could make disaster insurance products more affordable, could create strong incentives for consumers to buy insurance products, and would discourage unsustainable economic activities in disaster-prone areas.

While the previous articles dealt with the issue of adaptation to the direct consequences of climate change through insurance-based mechanisms, Axel Michaelowa examines the feasibility of using insurance-based mechanisms for offsetting the negative impacts of countries' adaptation measures in response to climate change. The necessity to address negative impacts of the implementation of mitigation and adaptation policies ('response measures') is specified in Articles 4.8 and 4.9 of the UNFCCC and Article 3.14 of the Kyoto Protocol.

By using a series of hypothetical but highly illustrative examples, the author demonstrates how adaptation policies of one country can adversely affect other economies. One example of such an adverse impact is a foreseen reduction in the demand for fossil fuels due to global adaptation measures which are likely to result in reduced world market prices for these fuels, and which arguably would lead to lower revenues for fossil-fuel-exporting countries. Alarmed by the potential adverse impact of global adaptation measures on their economies, for a long time OPEC countries have argued in the international climate negotiations that they should receive compensation for reduced export revenues.

Michaelowa attempts to find a risk-management solution to this problem. He begins by examining the applicability of insurance-based mechanism to managing the risk of adverse implications of adaptation measures on the economies of fossil-producing countries. After a careful examination of the problem, he concludes that the insurability of this risk is highly questionable due to the wide range of parameters that influence energy markets, which make it impossible to unambiguously separate the price and quantity effect caused by adaptation measures. In addition, as the timing of the adverse impact of adaptation measures can be easily predicted and the insured losses from such measures would be impossible to diversify (due to their systemic effect), insurers would be unable to offer insurance cover for such a risk.

An alternative approach to mitigating the impact of mitigation measures on oil prices may lie with the traditional commodity markets, where long-term price hedging contracts can be bought by countries at risk. However, due to the impossibility of teasing out the effect of mitigation measures from other factors that may reduce the price, tradable oil price hedging contracts are universal (e.g. cover against any cause of price decrease) and therefore relatively expensive.

The author concludes that the best long-term risk management policy for countries exporting fossil fuels is to diversify away from commodities in order to reduce the systemic market risk. Funds for diversification could be raised through taxes on the production of fossil fuels. These revenues could be used for investments in diversification projects, such as renewable energy technologies, which these countries can then export to offset their declining oil export revenues. This conclusion seems to be particularly sound in light of the fact that many fossil-fuel-exporting countries have a good renewable energy resource base in both solar and wind energy. Nevertheless, fossil-fuel exporters so far have neither taken up the opportunities of the pilot phase of Activities Implemented Jointly nor have they made visible efforts in the Clean Development Mechanism area.

Drawing on the material presented in this special issue, Eugene Gurenko concludes by drawing policy recommendations on how insurance-based mechanisms can best be utilized in the context of global adaptation to climate change. One of the key recommendations that also underpins every article in this Special Issue is that the creation of public–private partnerships in catastrophe insurance, where technical and capital resources of the insurance industry are combined with government actions to prevent and mitigate the risk of natural disasters, may be the only viable climate-adaptation strategy of the future.

Scientific and economic rationales for innovative climate insurance solutions

Peter Hoeppe[1]*, Eugene N. Gurenko[2]

[1]*Munich Re, Munich, Germany*
[2]*World Bank, Washington DC, US*

Abstract

The scientific and economic rationales for climate insurance solutions are provided in the context of global adaptation to climate change. Drawing on the growing body of scientific evidence on the increasing frequency and severity of climate-related natural disasters, we argue that climate change is already taking place. The mounting and highly unpredictable losses from natural disasters make the traditional disaster-funding approaches obsolete, as even large economies have problems financing economic recovery from their own budget revenues or special government disaster funds. This is particularly the case in low-income developing countries, where limited tax bases and high indebtedness prevent them from relying on debt financing of reconstruction efforts. Using OECD and World Bank statistics, we demonstrate that despite the commonly held belief, disaster-related external donor aid to developing countries accounts for only a small fraction of the total economic loss caused by catastrophic events. According to our estimate, on average over 90% of the economic loss from natural disasters is borne by households, businesses and government. This suggests a need for insurance-based climate risk financing mechanisms at the country level. By paying a fixed insurance premium that can be a small fraction of the potential economic loss, countries can cap the amount of their fiscal loss, greatly reduce the uncertainty of national budgetary outcomes due to natural disasters, and increase the speed of their post-disaster economic recovery.

Keywords: Adaptation; Climate change; Insurance; Natural catastrophes; Risk financing; Developing countries

1. Impact of climate change on global economic development

Over the last decades the frequency of major natural disasters as well as losses, both total economic and insured, caused by them have increased significantly. In Figure 1, it can be seen that over the last half-century (1950–2005), the frequency of 'great natural disasters' caused by different natural perils has been on the rise – from a global mean level of about two per year in the 1950s to about seven in recent years.

* Corresponding author. Tel.: +49-89-3891-2678; fax: +49-89-3891-72678
E-mail address: phoeppe@munichre.com

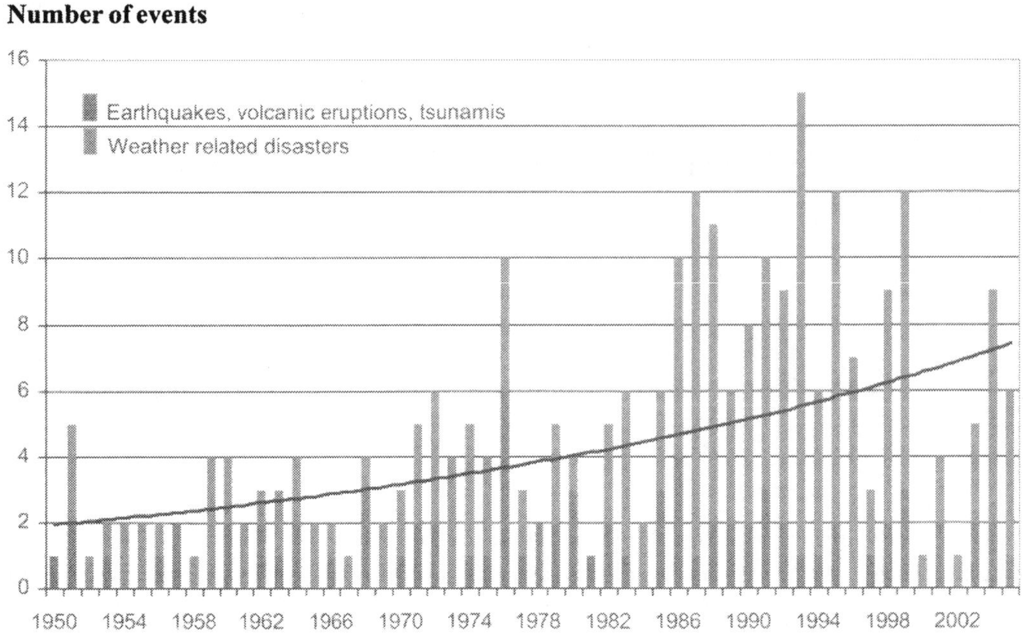

© 2006 NatCat*SERVICE,* Geo Risks Research, Munich Re
Figure 1. Great natural disasters, 1950–2005.

In this context, 'great natural disasters' are defined as events in which the affected region's ability to help itself is distinctly overtaxed. One or more of the following criteria apply:

- Interregional or international assistance is necessary
- Thousands are killed
- Hundreds of thousands are made homeless
- Substantial economic losses
- Considerable insured losses.

As great disasters are well documented in the newspapers and other media, there is little room for a reporting bias in these data. We are also quite convinced that the trend in the number of these great disasters, contrary to the level of economic damage caused by them, has no relevant confounding by population growth and increasing values. This means that a great disaster in 2004 would also have been a great disaster in 1950, even with less people involved and lower values affected in the latter case. Another interesting result from the data presented in Figure 1 is that there is no relevant trend for natural events of geophysical origin, such as earthquakes, volcanic eruptions or tsunamis (all represented by red bars). This means that the upward trend in the number of annual events is carried solely by weather-related events, which are inherently linked to climate change.

As can be seen from Figure 2, compared to the number of events, the trends in total economic and insured losses (all values already adjusted for inflation to values of 2005) are much more pronounced.

Figure 2 shows economic and insured losses only from great weather-related disasters. The economic losses in the last decade (1996–2005) have increased by a factor of seven as compared

Economic and insured losses

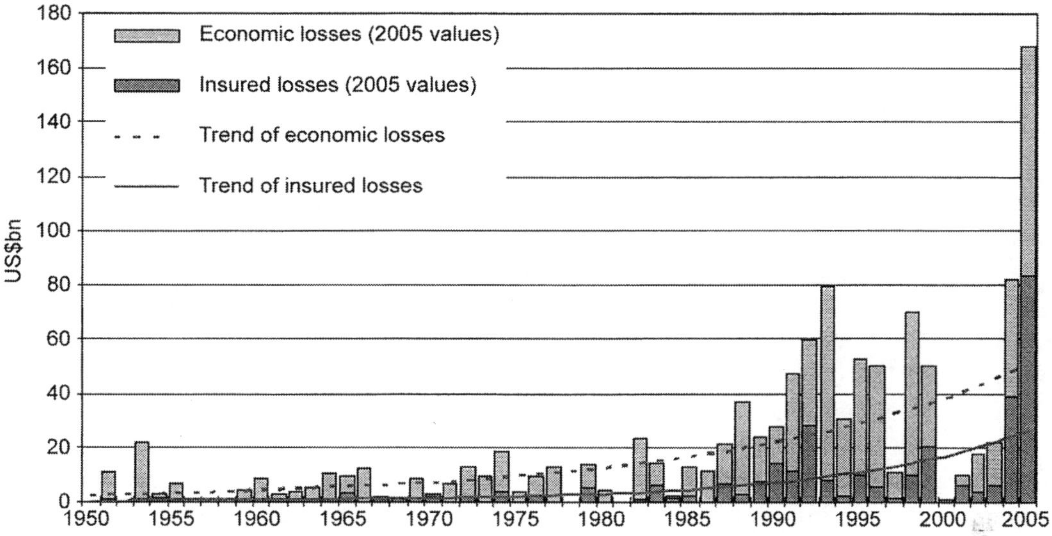

© 2006 NatCat*SERVICE*, Geo Risks Research, Munich Re

Figure 2. Development of economic and insured losses (in values of 2005) due to great weather-related disasters, 1950–2005.

with the 1960s level, and insured losses by a factor of 25. First 2004, and then 2005, have been the years with the highest-ever insured losses due to weather-related natural catastrophes. The trend of economic and insured losses is primarily attributable to the steady growth of the world population, the increasing concentration of people and economic value in urban areas, and the global migration of populations and industries into areas, such as coastal regions, that are particularly exposed to natural hazards. Yet, from the first results of an ongoing study of climate change by Munich Re, there seems to be a significant influence of climate change that can be seen not only through the increasing number of events, but also their atmospheric intensification.

During the last years there have been more and more indicators that climatic change is already influencing the frequency and intensity of natural catastrophes: e.g. the century flood in Saxony in 2002, the 450-year event of the extremely hot summer in Europe in 2003, and the all-time hurricane and typhoon record years of 2004 and 2005. In 2004, the first ever hurricane (Catarina) formed in the South Atlantic and caused significant damage in Brazil; in 2005 hurricane Vince formed close to the island of Madeira, the furthest northeast a tropical cyclone had ever developed in the Atlantic. Until recently, such phenomena had been thought to be impossible because of the relatively unfavourable conditions for the genesis of tropical storms there. The year 2005 has already set other records for hurricanes in the North Atlantic: never since the beginning of the records (1850) have so many devastating named tropical storms (seven by the end of July) developed that early in the season, and never before has a total number of 27 (including Zeta) been reached in one hurricane season (the previous record was 21). According to the World Meteorological Organization (WMO), the years 2001–2004 were among the five warmest recorded worldwide since 1856, with 2005 being the second warmest ever; which is yet more evidence of global warming.

Although the IPCC, in their 2001 report, still presented no clear proof of the correlation between global warming and the increased frequency and intensity of extreme atmospheric events, recent studies and simulations have provided a good deal of evidence that the probabilities of various meteorological parameters reaching extreme values are changing. A recent model simulation for the North Atlantic suggests that climate change will intensify the maximum wind speed by 0.5 on the Saffir–Simpson scale and precipitation by 18% in hurricanes until 2050 (Knutson and Tuleya, 2004). British scientists have estimated that it is very likely (confidence level >90%) that human influence has already at least doubled the risk of a heatwave exceeding the threshold magnitude of the European heatwave of 2003 (Stott et al., 2004). Recent publications by Emanuel (2005) and Webster et al. (2005) show for the first time that major tropical storms, both in the Atlantic and the Pacific region, have already increased since the 1970s in duration and intensity by about 50%. They predict that this trend induced by global warming will continue in the future. A study by Barnett et al. (2005) has demonstrated that the sea-surface temperatures in the areas relevant for tropical storms have already increased due to global warming by 0.5°C.

If the scientific global climate models tell us the truth, the present problems will be magnified in the near future. Changes in many atmospheric processes might significantly increase the frequency and severity of heatwaves, droughts, bush fires, tropical and extratropical cyclones, tornadoes, hailstorms, floods and storm surges in many parts of the world. These events will inevitably have a profound impact on property as well as also affecting the health and livelihood of many people.

We have to expect:

- increases in weather variability
- new extreme values for temperature, precipitation or wind speed in certain regions
- new exposures (like hurricanes in the South Atlantic)
- more frequent and devastating disasters.

The decisive question today is not when we will have the ultimate proof for anthropogenic climate change – a small risk of error will certainly still remain for some time – but which strategies we should follow to both mitigate and adapt to the change.

2. Access to insurance

One important step towards mitigating the effects of global warming is to provide proper insurance solutions to at least minimize the adverse financial consequences of an increasing number of natural catastrophes for countries and populations at risk. As shown in Figure 3, the worldwide distribution of insurance availability is very inhomogeneous. While the industrialized countries in North America, Europe and Australia enjoy a high level of insurance penetration, in Africa, Asia, and Latin America there are many countries with hardly any catastrophe insurance available.

The role played by commercial catastrophe insurance today in financing losses from natural disasters is explored further in Figure 4. The figure shows absolute annual insured losses on a 5-year-average basis for high income and lower income countries, as well as the trend line. A simple comparison of insured losses with overall economic losses from natural disasters (as depicted in Figure 7) shows a great disparity in the level of insurance coverage between the rich and the poor countries. While, in developed countries, the role of commercial disaster insurance

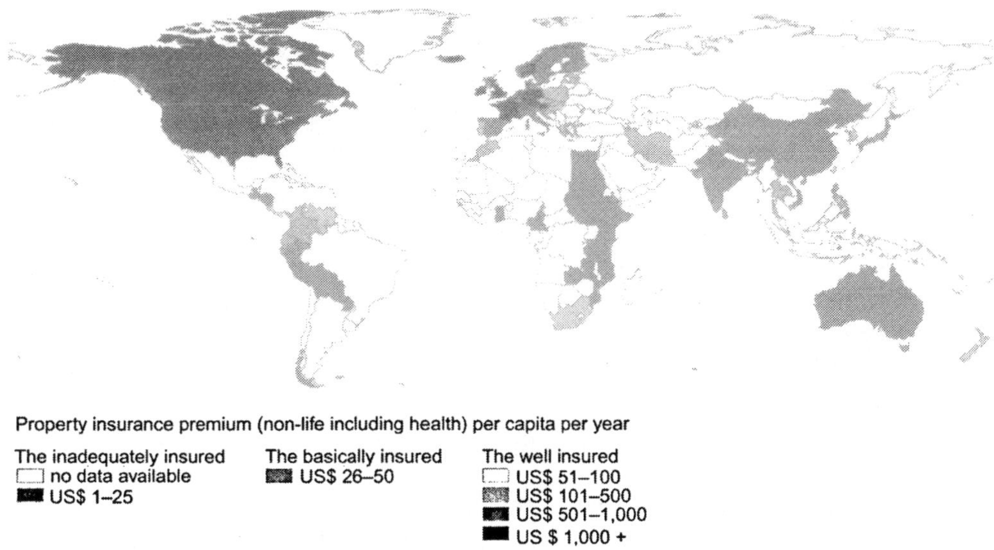

© 2006 Geo Risks Research, Munich Re
Figure 3. Global distribution of insurance premiums per capita.

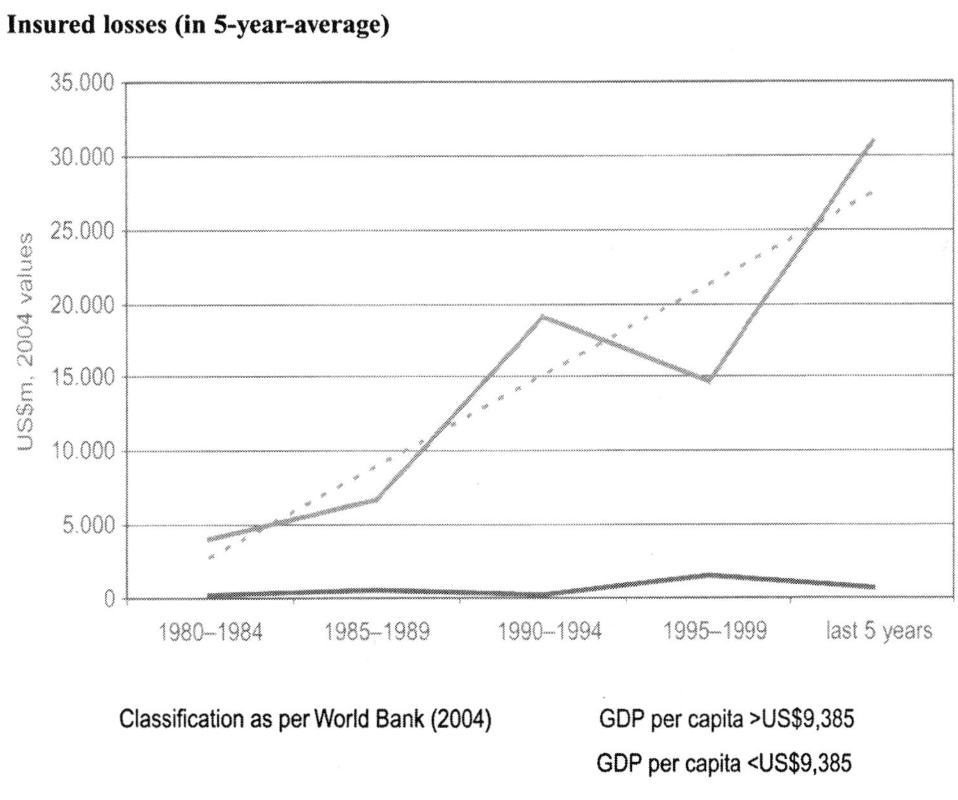

© 2005 Geo Risks Research, Munich Re
Figure 4. Insured losses from natural disasters in high-income versus middle/low-income countries.

in financing natural disasters has increased over the last 20 years from about 20% of economic loss in the early 1980s to about 40% today, the share of economic loss covered by insurance in developing countries has remained almost stagnant over the same period, accounting for about 3% of total economic loss. Although, to a large extent, such a disparity in insurance coverage can be explained by major differences in countries' levels of income and wealth, we must also point out the level of risk awareness, overall insurance culture, and finally, the extent to which private citizens are prepared to rely on governments for financial support in the aftermath of natural disasters.

An interesting question for the choice of regional scope and design of a climate insurance system is, whether there are differences between wealthy regions with an already high insurance density and other regions with little insurance availability in terms of their exposure and vulnerability to weather-related disasters. To answer this question some new analyses have been carried out at Munich Re. Figure 5 shows a map of the global distribution of great natural disasters between 1980 and 2005.

From Figure 5 one can hardly discern any difference in the pattern of natural disasters between 'wealthy' and 'poorer' countries. The USA, EU countries and Japan seem to be affected to a similar extent as the Caribbean States, India, the Philippines and China. In Figure 6, we explore the same question of potential differences in disaster patterns that may exist between four different income-groups of countries (in terms of GDP) intertemporally by looking at the annual number of weather-related catastrophes (all damaging events, not only great disasters).

By far the largest number of such events have occurred in the countries in the highest GDP class (>US$9,385), while between the other three classes there is hardly any difference. In all classes, however, there is a common upward trend in the number of annual events. Since the 1980s, the number of weather-related disasters increased from 180 events in the highest GDP class and about 50 events in the lower GDP classes to about 300 and 100 events, respectively, in 2004.

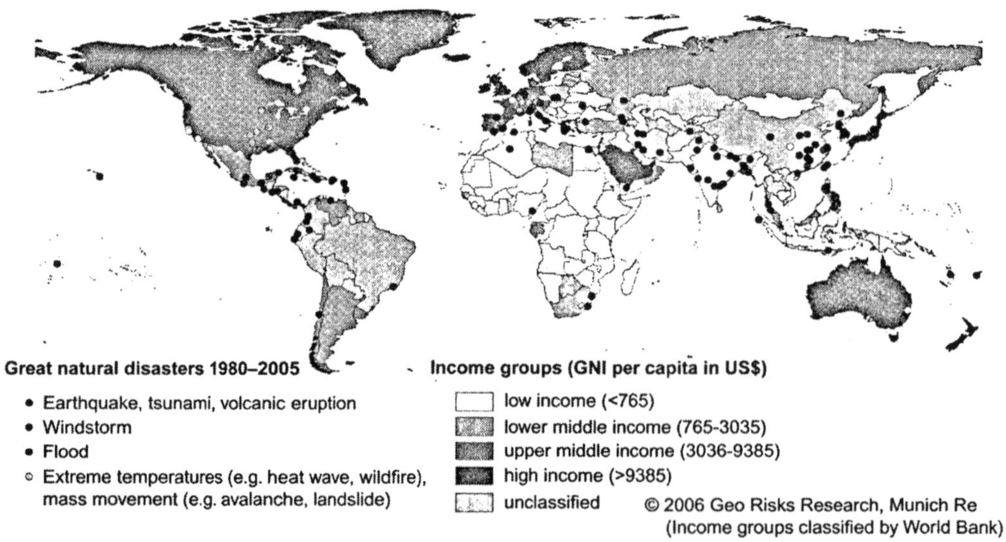

Figure 5. Natural catastrophes in economies at different stages of development between 1980 and 2005.

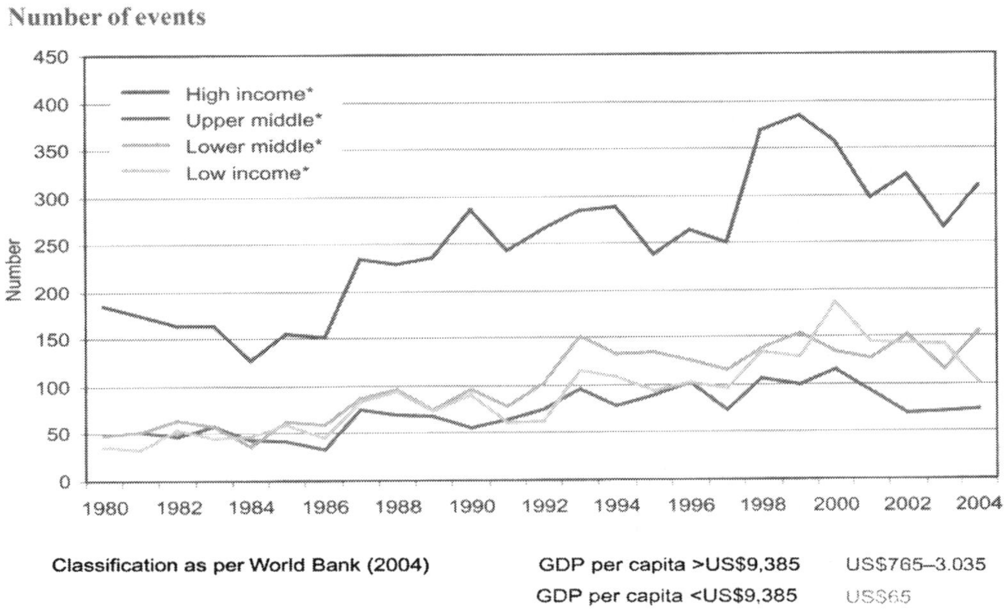

© 2005 Geo Risks Research, Munich Re

Figure 6. Weather-related catastrophes (1980–2004) in economies at different stages of development.

The total economic losses caused by weather-related disasters show a similar upward trend. In Figure 7, one can see a visible increase in economic losses caused by natural disasters, with the losses almost doubling for both 'high-income' and 'low/medium-income' groups of countries over the last 20 years. Due to the considerably higher concentration of values per area and a larger number of events, the high-income countries have experienced the highest absolute increases in economic losses from natural disasters – from about US$20 billion in the early 1980s to around US$70 billion in the early 2000s.

The absolute increase in economic losses for poorer countries looks more modest – from US$10 billion in the early 1980s to about US$15 billion in the early 2000s. However, if expressed as a percentage of GDP, economic losses caused by weather-related disasters for developing countries have been much more pronounced than those in industrialized countries. Between 1985 and 1999 alone, due to the considerably higher vulnerability of their economies to natural disasters, they lost 13.4% of their combined GDP versus only 2.5% in industrialized nations (Freeman and Scott, 2005).

As can be expected from the higher number of events and larger values at risk, the 'high-income' countries with GDP >US$9,385 experienced the largest intertemporal variation in economic losses. However, a high degree of variation in annual economic losses can also be seen among the 'middle/low income' class countries (GDP <US$9,385), pointing to the inadequacy of budget-based approaches to funding the highly variable disaster relief and reconstruction costs that prevail in developing countries (Gurenko and Lester, 2003). The growing volatility of economic losses experienced by all countries in the aftermath of natural disasters clearly demonstrates the rationale for risk financing instruments, such as insurance, that can smooth out this variability of outcomes for a fixed insurance premium.

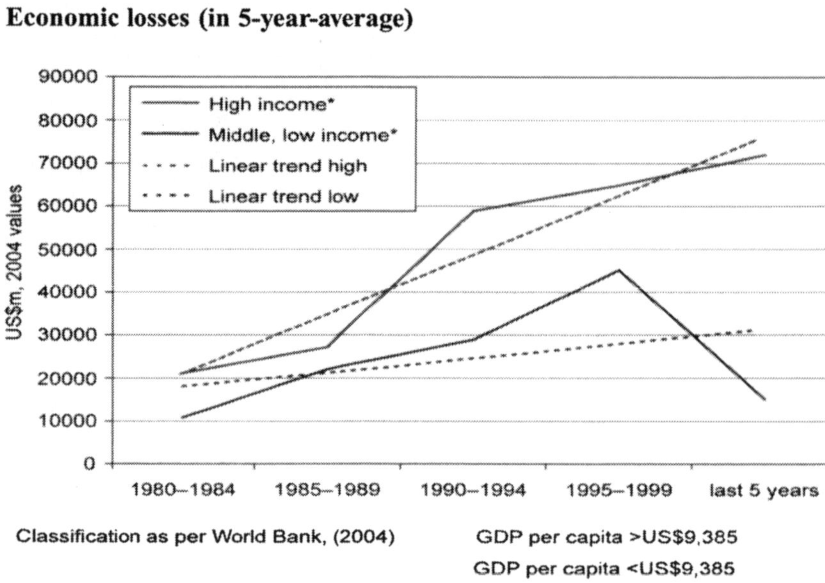

© 2005 Geo Risks Research, Munich Re
Figure 7. Development of economic losses caused by weather catastrophes 1980–2004 in economies at different stages of development.

3. Donor aid and development lending for natural disasters

A good measure of the capability of individual countries to cope with a natural disaster is the ratio of the economic damages caused by natural disasters to the GDP and the countries' fiscal resources (e.g. annual budgets), which are typically less than 50% of GDP.

In Table 1 some examples are given for countries in the Caribbean after the 2004 hurricane season. From these data it becomes clear that the economic damages caused to some countries by the hurricanes have been so severe that they could not recover without help from the international community.

As the frequency and scope of losses due to major natural catastrophes, especially tropical storms, is likely to be on the rise in the future, this example highlights the necessity for adaptation measures, including mitigation, and *ex-ante* risk financing solutions, including insurance, to enable these small disaster-prone nations to successfully recover from such devastating events.

Table 1. Hurricane losses in the selected Caribbean States in 2004 (GDP%)

Caribbean State	Losses compared to annual GDP
Dominican Republic	1.9%
Bahamas	10.5%
Jamaica	8.0%
Grenada	212.0%
Cayman Islands	183.0%

Source: Munich Re (2005) Geo Risks Research.

Up until now, most Caribbean countries have been relying on external concessional borrowings from international development banks (such as the World Bank, IDB and the IMF) and international donor aid to deal with the devastating consequences of natural disasters. In fact, reliance on these two sources of funding has been a major reason for the lack of insurance solutions for small-island States. However, there is clear evidence that over-reliance on these traditional post-disaster funding models may no longer be sustainable.

The increasing frequency and severity of natural disasters worldwide makes it more and more difficult for disaster-prone nations, particularly smaller sized economies, to finance economic losses in the aftermath of natural disasters out of recurrent or even future government budget revenues, due to the limited tax base and considerable indebtedness of many of these nations. As shown in Table 2, the level of indebtedness of small-island States in the Caribbean is about four times of that for middle-income countries, which means that the room for further borrowings to finance economic recovery efforts in the aftermath of future natural disasters is severely constrained.

Donor aid has been yet another major source of risk financing for most disaster-prone developing countries. Over-reliance on this source of funding, however, has major limitations. First, by its

Table 2. Indebtedness of selected CARICOM States

(Public and Publicly Guaranteed DOD as a % of GNI)				
	2001	2002	2003	Change 2000–03
Barbados	29%	29%	29%	7%
Belize	82%	93%	110%	39%
Dominica	79%	86%	89%	27%
Grenada	49%	78%	74%	26%
Guyana	168%	172%	175%	4%
Jamaica	56%	59%	60%	11%
St. Kitts and Nevis	71%	85%	103%	52%
St. Lucia	27%	33%	37%	10%
St. Vincent & the Grenadines	50%	51%	55%	3%
Trinidad and Tobago	20%	20%	17%	–6%
Average Small States				
Africa	125%	135%	127%	3%
Asia	41%	47%	44%	6%
Caribbean	63%	71%	75%	17%
All Small States	**82%**	**89%**	**86%**	**7%**
Memo:				
All developing countries	23%	23%	22%	–2%
Low income	36%	36%	34%	–5%
Lower middle income	25%	24%	21%	–6%
Upper middle income	15%	17%	17%	2%
Middle income	21%	21%	20%	–2%

Source: World Bank, January 2005.

very definition, donor aid is not a contractual obligation of donor governments and hence its delivery is subject to considerable political uncertainty. There is evidence that donor aid is more likely to be forthcoming in cases of highly catastrophic and internationally publicized events than in cases of more frequent but less devastating events, leaving considerable post-disaster funding gaps (Freeman et al., 2003).

In addition, as the amount of overall donor aid remains rather stable over time as a percentage of donor countries' GDP, which has been increasing in the order of 2–3% in the last decade, while economic losses caused by natural disasters have grown at a much more rapid pace, the ability of international donors to provide sufficient post-disaster financial assistance to disaster-prone nations in the future without reducing their financial commitment to other critical areas of economic development becomes a major problem.

As can be seen from Table 3, if in 1987–1989 the overall emergency and distress relief assistance accounted for only 1.6% of total donor assistance to developing countries, in 2003, it was 8.5% of total, or $5.87 billion. However, only about one-third of this assistance was earmarked for natural disasters, while the rest was used for complex emergencies (IMF, 2003). Taking this into account, the share of natural disaster aid in overall donor aid would account for only 1.3% and 4.3% in 1987–1989 and 2001, respectively. When expressed as a percentage of overall economic losses sustained by the developing countries, the donor assistance accounted for about 1% in 1987–1989 and about 9.6% in 2003. While illustrating the growing role of donor funding in financing economic losses caused by natural disasters in developing countries, these statistics mainly underscore the fact that donor funding is clearly insufficient to meet the growing disaster risk financing needs of developing economies. Given that insurance penetration in developing countries has been almost non-existent, most of the economic losses from natural disasters had to be absorbed by developing countries themselves.

Table 3. OECD development assistance statistics[a]

US$, millions	1987–1989 average	1990	1991	1992	1993	1994	1995	1996	1997	1998	1999	2000	2001	2002	2003
Economic losses from all natural events	24540	24790	37007	24384	43321	19424	45926	43711	27228	71967	40822	12071	16659	13084	20292
Emergency and distress relief aid	704	1058	2418	2586	3250	3468	3062	2963	2165	2787	4414	3574	3276	3869	5874
As percentage of ODA	1.61	2.1	4.2	4.4	5.9	5.8	5.2	5.3	4.5	5.5	8.5	7.2	6.5	6.64	8.51
Donor assistance for natural disasters as percent of economic losses[b]	0.9	1.4	2.2	3.5	2.5	5.9	2.2	2.2	2.6	1.3	3.6	9.8	6.5	9.8	9.6

Sources: OECD (2005), Munich Re Geo Risks Database for economic losses.

[a] Data also include allocations for post-conflict crises.

[b] Absolute amount of donor assistance for natural disasters was assumed to be one-third of total emergency and distress relief aid.

In Table 4, we provide annual estimates of the amount of economic loss from all natural disasters, including earthquakes and climate-related events, which had to be absorbed by developing countries over the last 17 years. We calculate it as an amount of overall economic loss caused by natural disasters less the donor assistance and insurance. For the sake of simplicity, we do not take into consideration emergency reconstruction loans provided by international development banks, as most of those would have to be eventually repaid and hence should be counted as a form of risk retention.

During 1987–1989, developing countries absorbed on average around 93% of total economic loss from natural disasters or about US$31 billion per year. If, in 1987–1989, developing countries retained on average around 95% of total economic loss from natural disasters or about US$23.3 billion, in 2003 their annual loss retention has decreased down to about 90% or over US$18.3 billion, mainly due to the increased share of donor funding allocated for natural disasters. Also, as can be seen from Table 4, the overall amount of losses from natural disasters absorbed by the developing countries is not only large but also highly variable, as measured by the coefficient of variation, which in this case is 50%.[1] Such loss volatility further exacerbates the level of social and economic disruptions caused by catastrophic events and points to the importance of insurance solutions. With the frequency and severity of natural disasters on the rise, it is obvious from these statistics that the existing model of financing natural disasters in developing countries is unlikely to be sustainable in the long run, due to the increasing volatility of global climate and the growing resource gap between the overall economic damages sustained by developing countries and the available financial assistance from the donors and commercial insurers to finance them.

A part of the above mentioned funding gap caused by natural disasters can be covered by concessional lending from development banks, such as the World Bank, Inter-American Development Bank, and Asian Development Bank. In fact, loans for disaster reconstruction purposes have become an important part of their lending portfolios. As can be seen in Figure 8, since the early 1980s the World Bank's lending for disaster reconstruction purposes has been on the rise, with much of this lending being quite recent. All in all, during this period the World Bank has originated 528 loans that, in one way or another, addressed the risk of natural disasters. Yet, similar

Table 4. Economic losses from natural disasters retained by developing countries, 1987–2003

US$ millions	1987–1989 average	1990	1991	1992	1993	1994	1995	1996	1997	1998	1999	2000	2001	2002	2003
Economic losses from all natural events	24540	24790	37007	24384	43321	19424	45926	43711	27228	71967	40822	12071	16659	13084	20292
Emergency and distress relief aid[a]	232	349	797	853	1072	1144	1010	977	714	919	1456	1179	1081	1276	1938
Insured loss	972	537	1022	47	103	230	532	581	1210	4688	1703	143	886	1646	35
Total retained loss	23335	23903	35187	23483	42145	18050	44384	42152	25303	66359	37662	10749	14691	10161	18318
Retained loss as percentage of total loss	95.1	96.4	95.1	96.3	97.3	92.9	96.6	96.4	92.9	92.2	92.3	89.0	88.2	77.7	90.3

Sources: IMF Working Paper (2003), OECD (2005), Munich Re Geo Risks Database for economic and insured losses.
[a] Absolute amount of donor assistance for natural disasters is assumed to be one-third of total emergency and distress relief aid.

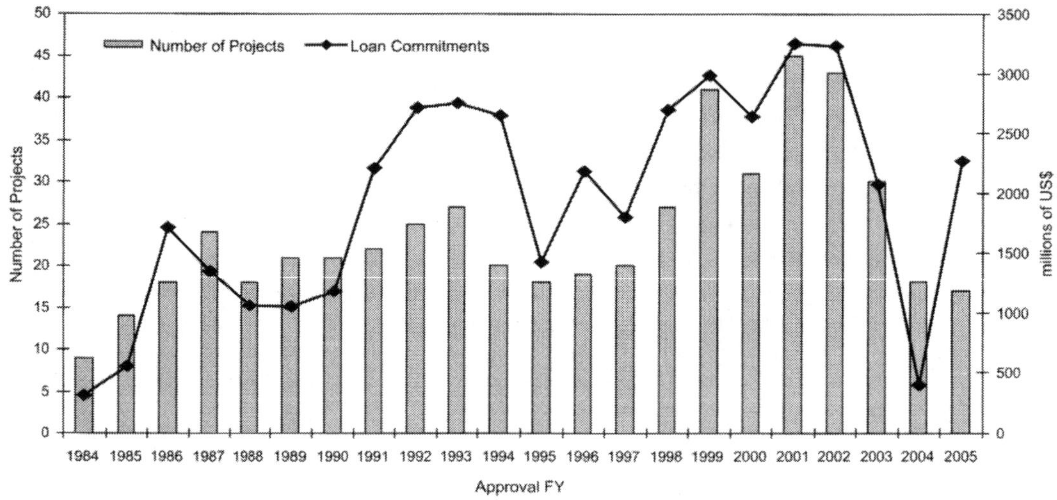

Figure 8. World Bank lending for natural disasters, 1984–2005.

to the donor aid, most of this lending has been provided in the aftermath of natural disasters and carried few incentives for countries to engage in proactive risk management. In addition, despite the growing percentage of the Bank's lending allocated to natural disasters, the amount of reconstruction lending has been small relative to overall economic losses – on average about 2% of economic loss retained by developing countries. Another important drawback of reconstruction loans is that it typically takes up to 1 year for them to disburse, which leaves governments scrambling for liquidity in the first few months after a disaster.

Last but not least, the existing *ex-post, ad-hoc* model of financing natural disasters losses in developing countries that has been widely adopted by the international donor community and development lenders fails to provide disaster-prone countries with sufficient incentives for mitigation and risk reduction. As donor funding arrives in the aftermath of major catastrophic events, and by and large is used for emergency relief and reconstruction purposes, very little of this aid is invested in mitigation projects to reduce losses from similar catastrophe events in the future. As opposed to commercial property insurers, which frequently link the very availability of insurance coverage to the implementation of concrete risk reduction measures by the insured, donors require nothing of that sort from disaster-prone countries. As a result, countries at risk see little economic or political benefit from investing in mitigation or better enforcement of construction codes or land-use policies that would restrict construction activities in harm's way. The unfortunate outcome of these disaster funding policies can be seen clearly in Figure 9.

Despite the overall focus of this article on the economic implications of weather-related hazards, we thought the example of seismic vulnerability of structures in developing countries would provide a useful illustration of the matter at hand. Figure 9 depicts aggregate seismic composite vulnerability curves for residential and commercial structures in developing (LD) and highly developed (HD) economies. Vulnerability is measured in terms of the mean damage factor, which is the ratio of the cost of repair to the total insured value. Vulnerability functions are defined in terms of the type of the structural system (for example, frame or walls), the method and time of construction, and the construction material. Typically, they are developed on the basis of an analysis of claims data from catastrophe events throughout the world, engineering-based analytical studies, expert opinion and laboratory tests.

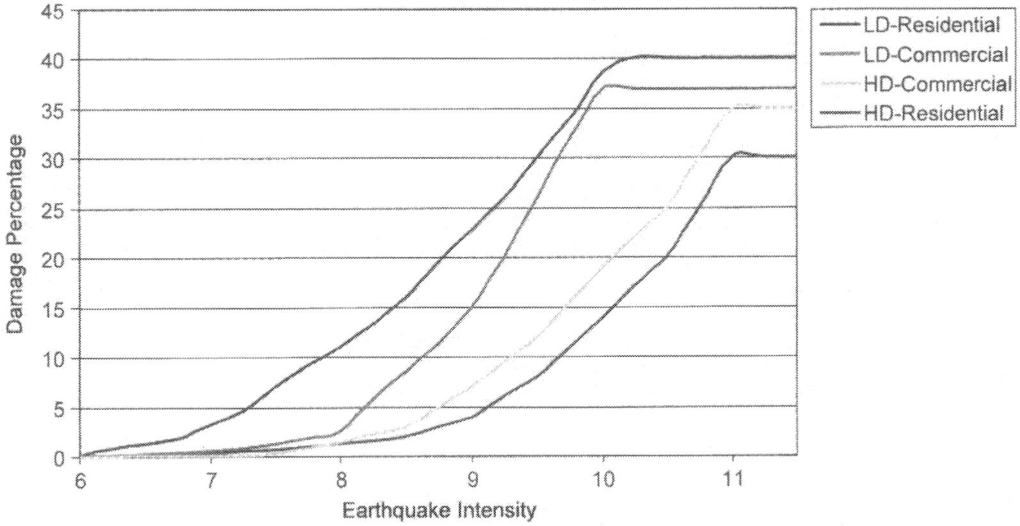

© 2005 Geo Risks Research, Munich Re

Figure 9. Seismic vulnerability of commercial and residential construction in developing (LD) and developed (HD) countries.

Figure 9 shows that buildings in developing countries are much more vulnerable than in highly developed countries, whose construction standards and enforcement of building codes are stronger. For instance, an earthquake of intensity 9 would cause a mean damage of 23% for residential and 15% for commercial buildings in developing countries, while the same event would cause a mean damage of less than 7% for commercial and 4% for residential structures, respectively, in highly developed countries. These statistics demonstrate the urgency of developing more effective risk financing policies by the donor community that can encourage developing countries to invest in reducing their vulnerability to natural disasters in the future. By reducing the physical vulnerability of structures to natural hazards and investing in risk mitigation projects, developing countries will not only save billions of dollars in future economic losses but, more importantly, save thousands of lives.

4. Conclusions

The number of weather-related disasters and the economic losses caused by them have been rising during the last decades and will continue to do so in the future due to climate change. Although the economic losses caused by natural disasters are the highest in industrialized countries, in relative terms their overall impact on these economies has been rather minimal, as they still have sufficient financial and technological resources to absorb it. However, for many of the poorer countries, the increasing exposure to natural catastrophes in conjunction with the higher vulnerability of their economies to natural disasters and highly volatile and insufficient external financial assistance entails large risks for their economic and social development. In the absence of new innovative global catastrophe risk financing mechanisms, including catastrophe insurance, that can address the increasing volatility and severity of losses sustained by these economies due to natural disasters, and, at the same time, provide appropriate incentives for *ex-ante* risk management for disaster-prone countries and their populations, the adverse impact of the global climate change is likely to become even more pronounced in the future.

Note

1 A coefficient of variation is a ratio of variable's standard deviation to the mean.

References

Barnett, T.P., Pierce, D.W., Achutarao, K., Gleckler, P., Santer, B., Gregory, J., Washington, W., 2005. Penetration of human-induced warming into the world's oceans. Science 8 July, 284–287 [published online 2 June 2005, doi:10.1126/science.1112418].
Emanuel, K., 2005. Increasing destructiveness of tropical cyclones over the past 30 years. Nature 436, 686–688.
Freeman, P., Scott, K., 2005. Comparative analysis of large scale catastrophic compensation schemes. In: Catastrophic Risks and Insurance. OECD Vol. 8. OECD Publishing.
Freeman, P., Martin, L., Linnerooth-Bayer, J., Mechler, R., Pflug, G., Warner, K., 2003. Disaster Risk Management. Washington, DC, Sustainable Development Department of the Inter-American Development Bank.
Gurenko, E., Lester, R., 2003. Financing the Risk of Natural Disasters in India: A Risk Management Framework. World Bank Report.
IMF, 2003. Fund Assistance for Countries Facing Exogenous Shocks. Working Paper, IMF, August 2003.
IPCC, 2001. Climate Change 2001: Impacts, Adaptation and Vulnerability. Cambridge University Press, Cambridge, UK.
Knutson, T.R., Tuleya, R.E., 2004. Impact of CO_2-induced warming on simulated hurricane intensity and precipitation: sensitivity to the choice of climate model and convective parameterization. Journal of Climate 17, 3477–3495.
Munich Reinsurance Company, 2006. Annual Review of Natural Disasters 2005. Topics, Munich Re, Munich, Germany.
OECD, 2005. DAC Tables.
Stott, P.A., Stone, D.A., Allen, M.R., 2004. Human contribution to the European heatwave of 2003. Nature 432, 610–614.
Webster, P.J., Holland, G.J., Curry, J.A., Chang, H.-R., 2005. Changes in tropical cyclone number, duration, and intensity in a warming environment. Science 309, 1844–1846.

Insurance for assisting adaptation to climate change in developing countries: a proposed strategy

Joanne Linnerooth-Bayer*, Reinhard Mechler

Risk and Vulnerability Programme, International Institute for Applied Systems Analysis, Laxenburg, Austria

Abstract

This paper suggests a two-tiered climate insurance strategy that would support developing country adaptation to the risks of climate variability and meet the intent of Article 4.8 of the United Nations Framework Convention on Climate Change (UNFCCC). The core of this strategy is the establishment of a climate insurance programme specialized in supporting developing country insurance-related initiatives for sudden- and slow-onset weather-related disasters. This programme could take many institutional forms, including an independent facility, a facility in partnership with other institutions of the donor community, or as part of a multi-purpose disaster management facility operated outside of the climate regime. Its main purpose would be to enable the establishment of public–private safety nets for climate-related shocks by assisting the development of (sometimes novel) insurance-related instruments that are affordable to the poor and coupled with actions and incentives for pro-active preventive measures. A second tier could provide disaster relief contingent on countries making credible efforts to manage their risks. Since it would be based on precedents of donor-supported insurance systems in developing countries, a main advantage of this proposed climate insurance strategy is its demonstrated feasibility. Other advantages include its potential for linking with related donor initiatives, providing incentives for loss reduction and targeting the most vulnerable. Many details and issues are left unresolved, and it is hoped that this suggested strategy will facilitate needed discussion on practical options for supporting adaptation to climate change in developing countries.

Keywords: Climate change; Insurance; Disasters; Article 4.8; Adaptation; Vulnerability; Developing countries

1. Introduction

Adaptation to climate change, including support for insurance instruments, has emerged on the climate agenda alongside the reduction of atmospheric greenhouse gas concentrations as an essential part of the response to climate change risks. It is generally accepted that industrialized countries bear a certain responsibility for adaptation to climate change in developing countries, and should bear part of the costs. Although a diversity of mechanisms, approaches and rules for funding adaptation in developing countries has been adopted by implementing agencies and

* Corresponding author. Tel.: +43-2236-807308; fax: +43-2236-71313
E-mail address: bayer@iiasa.ac.at

governments in the context of the United Nations Framework Convention on Climate Change (UNFCCC), adaptation is generally considered to be an underdeveloped part of the climate regime. Early efforts to address adaptation under the UNFCCC have focused on capacity building and information exchange with respect to vulnerability to climate change and possible adaptation strategies. There is an increasing appreciation that minimizing vulnerability to the economic and physical impacts of climate-related extreme weather events, including floods, droughts, typhoons and other weather hazards, can cost-effectively contribute to reduced vulnerability. Moreover, many in the climate community are advocating that climate risk reduction be mainstreamed into the development process to simultaneously contribute to eradicating poverty, furthering development, and achieving the Millennium Development Goals. Efforts have aimed at funding strictly climate-*change*-related activities, but there are increasing calls that adaptation should be driven by vulnerability and poverty, and that it should be mainstreamed into the development process (Kartha et al., 2006).

The reduction of the escalating losses from floods, droughts, typhoons and other climate-related disasters is viewed as essential to eradicating poverty and achieving the Millennium Development Goals (Arnold and Kreimer, 2004). In the past quarter-century, over 95% of disaster deaths occurred in developing countries, and direct economic losses (averaging US$54 billion per annum) as a share of national income were more than double in low-income than in high-income countries (Arnold and Kreimer, 2004). Although the increase in disaster losses today is largely driven by socio-economic factors, there is mounting evidence of a significant climate-change signal in disaster events, for example, increasing extreme precipitation at mid- and high latitudes (Schönwiese et al., 2003), extreme floods and droughts in temperate and tropical Asia, severe dry events in the Sahel and southern Africa (IPCC, 2001), and increases in tropical cyclone activity in the Atlantic and the Pacific region (Emanuel, 2005). Scientists, however, cannot accurately assess the contribution of climate change to current risks. Nevertheless, many in the climate community are calling for a 'no-regrets' adaptation strategy that integrates adaptation to climate change with adaptation to 'normal' climate variability. Improving the capacity of communities, governments or regions to deal with climate variability will be likely to improve their resilience to deal with future climatic changes. This means that increasing attention must be paid to disaster risk management.

An important cornerstone for risk management, and a possible no-regrets adaptation strategy, is insurance and alternative risk-transfer instruments that provide disaster safety nets for the most vulnerable (Linnerooth-Bayer et al., 2005). Without donor support, however, insurance is hardly affordable in highly exposed developing countries, which helps to explain why only 1% of households and businesses in low-income countries, and only 3% in middle-income countries, have catastrophe coverage, compared with 30% in high-income countries (Munich Re, 2005). Instead of insurance, they rely on support from family and governments, which is not always forthcoming for catastrophes that affect whole regions or countries. Disasters exacerbate poverty as victims take out high-interest loans, sell assets and livestock, or engage in low-risk, low-yield farming to lessen their exposure to extreme events. Moreover, without a post-disaster infusion of capital for reconstruction, disasters can have long-term adverse effects on economic development. As a case in point, 4 years after the devastation of Hurricane Mitch in 1998, the GDP of Honduras was 6% below pre-disaster projections (Mechler, 2004), and the disaster increased the number of the poor by 165,000 people (Government of Honduras, 2001).

Climate risk management, including proactive support for insurance instruments, is emerging on the climate change adaptation agenda. Article 4.8 of the UNFCCC calls upon Convention Parties to consider actions, *including insurance*, to meet the specific needs and concerns of developing countries arising from the adverse impacts of climate change (United Nations, 1992), and Article 3.14 of the Kyoto Protocol explicitly calls for consideration of the establishment of insurance (United Nations, 1997). In an early proposal for an 'international insurance pool' within the UNFCCC context, the Alliance of Small Island States (AOSIS) put forth the idea of a global compensation fund fully financed by industrialized countries for the purpose of compensating low-lying states for sea-level rise damages. The AOSIS proposal addressed what is arguably an uninsurable risk (since sea-level rise is gradual and its occurrence predictable) for which the victims have little responsibility.

This article addresses a different risk context, that of stochastic sudden- and slow-onset weather-related disasters, and suggests a two-tiered climate insurance strategy. The first tier, and the core of this strategy, is the establishment of a climate insurance programme that would offer capacity building and financial support to nascent (weather) disaster insurance systems in highly exposed developing countries. This support could be offered independently or in partnership with other donor organizations by creating a climate insurance facility or other mechanism. Alternatively, it could be 'mainstreamed' into the operations of a multi-purpose disaster risk management facility. A main purpose of the climate insurance programme is to enable the establishment of public/private safety nets for stochastic climate-related shocks by assisting the development of insurance-related instruments that are affordable to the poor, coupled with actions and incentives for proactive preventive (adaptation) measures. As a second tier of support, adaptation funding could be apportioned to post-event relief for weather-related disaster risks that are otherwise uninsured because of data or institutional limitations.

The intent of this discussion is *not* to provide a concrete proposal for negotiation, but rather to suggest a broadly conceived climate insurance strategy as a basis for further discussion and deliberation. We begin in the next section by briefly reviewing the AOSIS and other recent climate insurance proposals that provide the background for our suggested strategy. We continue in Section 3 by outlining the workings of the first-tier climate insurance programme, which builds on developing country initiatives and thus avoids the expense and obstacles of operating an independent system. Based on experience in India, Malawi, Turkey and Mexico, we give concrete examples of the types of insurance initiatives that the programme might support. In Section 4, we offer preliminary thoughts on a possible second tier, which would provide disaster relief contingent on credible risk management policies or actions. Section 5 discusses challenges and opportunities for financing and implementing this two-tiered strategy. Section 6 concludes by briefly reviewing the advantages of this proposal, including its feasibility and potential for linking with other donor initiatives, providing incentives for loss reduction (adaptation) and targeting the most vulnerable. The unresolved issues are discussed, including the necessary institutional design, possible limits on support (for instance that funds be commensurate with the incremental risk of climate change), and sources for the requisite resources.

2. Climate insurance proposals

2.1. AOSIS proposal

Introducing the term 'insurance' for the first time, the Alliance of Small Island States (AOSIS) suggested in 1991 that an 'international insurance pool' funded by industrialized (Annex II) countries be established under the control of the Conference of the Parties (COP) to compensate small-island and low-lying developing nations for the uninsured loss and damage from slow-onset sea-level rise. The pool would

> compensate developing countries (i) in situations where selecting the least climate sensitive development option involves incurring additional expense and (ii) where insurance is not available for damage resulting from climate change (Intergovernmental Negotiating Committee, 1991).

Mandatory contributions to the fund would be made to an administrating authority, which would also be responsible for handling claims made against the resources of the fund. As a basis for settling claims, the proposal contemplated that assets in developing countries potentially affected by sea-level rise would be valued and registered with the authority. Trigger levels (levels of sea-level rise that would legally require the payment of claims) would be subject to negotiation between individual countries and the authority. Importantly, in assessing claims, the authority was to determine whether and to what extent the loss or damage could have been avoided by measures which might reasonably have been taken at an earlier stage, thus avoiding the moral hazard of not taking appropriate preventive measures. Assets covered by commercial insurance would not be compensated by the scheme.

There are difficult challenges in implementing the AOSIS proposal. Valuing all properties and verifying loss claims in countries with no indigenous insurance structures would impose large transaction costs on the system. Determining 'reasonable' loss-reduction measures is also problematic. Nonetheless, the proposal was, and remains, a valuable first step in presenting concrete ideas on how developed countries could take financial responsibility for climate-change impacts accruing to vulnerable developing countries.

2.2. Müller proposal

Whereas the AOSIS insurance proposal addressed the gradual onset of sea-level rise, subsequent proposals have turned to sudden-onset weather events such as floods, tropical cyclones and sea surges (worsened by sea-level rise). Müller (2002) advocated a switch from the current international disaster relief system characterized by voluntary, media-driven and uncoordinated donations to a Climate Impact Relief Fund (CIRF), which is regularly funded up-front and centrally administered by the UNFCCC in order to increase efficiency and fairness. No 'new money' would be needed, since OECD or Annex II countries would donate to the fund proportionally to their current average post-disaster assistance spending. According to Müller, further options for such a fund could be to provide disaster preparedness support and adopt burden-sharing criteria, such as based on financial ability or a CO_2-emission-based system.

2.3. Germanwatch proposal

The Germanwatch proposal for a Climate Change Funding Mechanism (Bals et al., 2006) builds strongly on the AOSIS and Müller proposals. The authors propose a global catastrophe insurance programme funded by developed countries and administered by a public/private entity. The scheme would be limited in scope by indemnifying only public infrastructure damage in least-developed countries (LDCs) and offering cover only for rare, high-consequence, climate-related risks. As an interesting innovation, there would be in-kind premium payments in the form of implemented loss-reduction measures by public clients who voluntarily join the scheme: the CCFM would define minimum risk reduction measures to be undertaken by the country where the annual cost to the country is commensurate with the level of imputed risk-based premium.

Defining risk-reduction measures by an outside authority (for example, requiring squatters to evacuate areas targeted for flood-control dams) may be problematic, especially if not subject to government and stakeholder involvement. Moreover, least-developed countries may find it difficult to finance mitigation measures that cover the imputed risk-based premium. For highly exposed LDCs, this premium can be quite substantial. For example, in the recently introduced drought insurance programme in Malawi, annual premiums amounted to 6–10% of the insured crop value (Opportunity International, 2005). Finally, the strategy can be inefficient if the required measures are not cost-effective or high priority in the country.

The Germanwatch strategy also faces problems in its practical implementation. Besides costly monitoring of adaptation measures, post-disaster losses must be assessed to determine the triggering threshold. This will involve high transaction costs, especially in the less-developed countries lacking insurance infrastructure and claims handling expertise. It will also encourage overestimation of loss figures, which will be difficult to verify. Assessing risks, setting in-kind premiums, monitoring adaptation measures and settling claims will require a large administrative apparatus. Finally, targeting governments for claims payments poses the same problem that donors confront with post-disaster aid – payments in the hands of corrupt officials may not reach their intended purpose. Despite the drawbacks, the Germanwatch proposal and its predecessors have strong merits. They target the most vulnerable and encourage proactive risk management measures in highly exposed countries.

3. Towards a complementary strategy for implementing Article 4.8

In a background paper prepared for a UNFCCC meeting on climate change and financial adaptation (Linnerooth-Bayer et al., 2003; see also Linnerooth-Bayer and Mechler, 2003) the authors suggest that implementation of Article 4.8 could be based on developed (Annex II) country support for developing country insurance initiatives. In this article, we elaborate on this earlier concept by proposing a two-tiered climate insurance strategy. As shown in Figure 1, the first tier would take the form of a climate insurance programme that provides support to nascent (climate-related) disaster insurance systems in highly exposed developing countries. The second tier would provide post-disaster relief to countries that demonstrate credible efforts in managing their risks. In this section we elaborate on the first tier of support.

Climate Insurance Strategy	
Tier 1: Climate Insurance Programme	**Tier 2: Disaster Response Programme**
Support for nascent disaster risk financing mechanisms for climate-related risks at the local and national levels	Relief for: • Climate-related disasters not (yet) financially protected • Uninsurable climate-related disasters

Figure 1. The two tiers of a climate insurance strategy.

In contrast to the Germanwatch proposal, which advocates the creation of a global insurance scheme with full responsibility on the relevant authority for underwriting risks and administering an insurance system, the first tier of this strategy would be based on shared responsibility at the local, national and global levels. The climate insurance programme could stand alone, for example, with the creation of an independent climate insurance facility, or it could operate in partnership with other organizations, including international financial institutions, bilateral donors, international organizations, non-governmental organizations and the insurance industry. Alternatively, the funds could be mainstreamed into a multi-purpose, multi-donor disaster risk management facility.

A main aim of this proposed climate insurance programme is to enable the establishment of public/private safety nets for stochastic weather-related shocks by making use of insurance instruments that are affordable to vulnerable and marginalized communities, coupled with actions and incentives for proactive preventive (adaptation) measures. As illustrated in Figure 2, this programme would provide assistance to a wide range of insurance-related initiatives, including schemes providing cover for (1) property, crops, life and health impacts, and (2) government liabilities for public infrastructure damages and relief spending. Assistance could take many forms, including technical support for feasibility studies and capacity building, and financial support in the form of reinsurance and subsidies. It could be extended to schemes at the local, national, regional and even global levels, complementing each other and leading to better global risk diversification and, as a consequence, reduced premiums.

Without this assistance, insurance programmes will not be viable in many highly exposed developing countries. Because of the high costs of insuring correlated or covariant disaster risks (which affect whole regions at the same time), individuals can pay substantially more than the

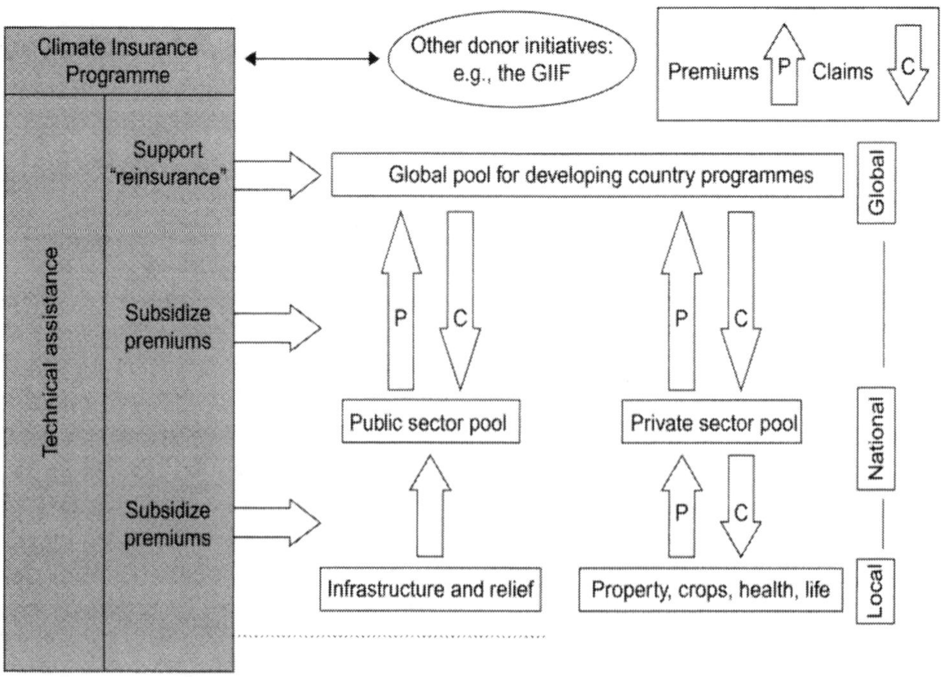

Figure 2. An illustration of the climate insurance programme.

expected losses they will experience over the long term, which may not be feasible or desirable without donor support. Donors can also ensure the proper design of insurance contracts to reward risk-reducing behaviour and thus avoid 'moral hazard', which means that individuals take fewer precautionary measures because they are insured (Brown and Churchill, 2000). Moreover, donors can promote the development of local catastrophe insurance markets by offering additional fairly priced reinsurance capacity. Such an approach will help reduce the risk of insurer insolvency and defaults on claims in the case of large or repeated catastrophes (Brown et al., 2000), and will contribute to making these systems accessible and affordable to the poor.

Four cases of recent donor-supported insurance initiatives, with illustrative examples, are described below and serve to illustrate the possibilities for a climate insurance programme.

3.1. Assisting index-based insurance for crops and livelihoods

More than 40% of farmers in developing countries face threats to their livelihoods from adverse weather (World Bank, 2005a). Weather risk destabilizes households and countries and creates food insecurity. In the Southern African Development Community (SADC), as a case in point, floods, cyclones and droughts have been a major cause of hunger affecting more than 30 million people since 2000. Governments and donors react to these shocks rather than proactively managing the risks. These emergency reactions have been criticized for being *ad hoc*, sometimes untimely, and destabilizing local food markets (Hess and Syroka, 2005).

Novel insurance instruments are emerging to address problems of food insecurity, even for high-frequency, slower onset disasters, such as droughts. Affordable insurance can provide low-income farm households with access to post-disaster liquidity, thus securing their livelihoods and avoiding famine. Moreover, insurance improves their credit worthiness and allows smallholder farmers to engage in higher-return crop practices. According to the World Food Programme (2005, p. 7):

> Because of the extreme and covariant nature of the risks they face, and in the absence of risk-management instruments such as crop insurance, risk-averse smallholder farmers naturally seek to minimize their exposure ... by opting for lower-value (lower-risk) and therefore lower-return crops, using little or no fertilizer and over-diversifying their income sources. These risk-management choices also keep farmers from taking advantage of profitable opportunities; they are a fundamental cause of continued poverty.

3.1.1. Example: Index-based insurance in Malawi

In Malawi, where the economy and livelihoods are severely affected by rainfall risk, resulting in drought and food insecurity, groundnut farmers can now receive loans that are insured against default with an index-based weather derivative (Hess and Syroka, 2005). This is a contingent contract with a payoff determined by weather events, in this case a specified lack of precipitation recorded at a specified weather station. Farmers collect an insurance payment if the index reaches a certain measure or 'trigger', regardless of actual losses.

The Malawi pilot project offers a packaged loan and index-based microinsurance product to groups of groundnut farmers organized by the National Smallholder Farmers Association. Accordingly, the farmer enters into a loan agreement with a higher interest rate that includes the weather insurance premium, which the bank and rural finance institution pay to the insurer, the Insurance Association of Malawi. In the event of a severe drought (as measured by the rainfall index), the borrower pays only a fraction of the loan due, and the rest is paid by the insurer directly to the bank. Without this insurance, banks rarely loan to high-risk, low-income farmers, which means they cannot obtain needed credit to invest in the seeds and other inputs necessary for higher-yield crops. Moreover, because of the physical trigger, there is no moral hazard; on the contrary, farmers will have an incentive to reduce potential losses, for example, by diversifying their crops. Nor is there a need for expensive individual claims-settling, and expedient payments will reduce the need for farmers to sell their assets and livestock to survive the aftermath of a disaster. One drawback of index insurance, however, is 'basis risk', which means that payouts may not be fully correlated with losses.

The World Bank has provided technical assistance and training in developing this weather insurance product (H. Ibarra, personal communication, 2005). By reducing loan repayments in the case of drought, the Malawi scheme only indirectly protects farmers from loss of livelihood and food insecurity. Still, it illustrates a large potential for donor-supported, index-based schemes that can, in contrast to the Malawi case, be designed to provide needed liquidity after major disasters. In India, for example, international technical assistance has been instrumental in the current success of index-based crop insurance programmes, which have increased penetration from 230 farmers to over 250,000 over a 3-year period, and similar schemes have been implemented or are under way in Mongolia, Ukraine, Peru, Thailand and Ethiopia (Mechler et al., 2006). Unless supported by technical assistance, national subsidies (cross-subsidies, as in India), or international donors, these schemes are out of reach for very low-income smallholder farmers. As illustrated in Figure 2, providing this support presents an opportunity for a climate insurance programme.

3.2. Assisting microinsurance schemes for property and life in low-income countries

Any discussion on how to support risk pooling and transfer in developing countries must take into account the capacity to engage in such efforts by the vulnerable and marginalized. The alternatives to insurance for many in the developing world include arrangements that involve reciprocal exchange, such as kinship ties and community self-help. Despite their limitations, Cohen and Sebstad (2003) claim that these risk-sharing arrangements work reasonably well for less severe and idiosyncratic shocks. However, they are inadequate and inappropriate for catastrophes that affect people throughout a region or country. Without reciprocal support or outside aid, disasters can lead to a 'cycle of poverty', as victims take out high-interest loans (or default on existing loans), sell assets and livestock, or engage in low-risk, low-yield farming to lessen their exposure to extreme events (Siegel, 2005).

3.2.1. Example: Microinsurance in India

The coastal Andhra Pradesh region of India is exposed to multiple and severe hazards, including floods, landslides, earthquakes and cyclones. Since 2004, microinsurance services have been provided in this region as part of the voluntary Disaster Preparedness Programme, which also offers capacity building of communities, government, civil society and media organizations. In partnership with the Oriental Insurance Company, this programme offers multiple-hazard insurance coverage for property and life risks to groups of women with a minimum size of 250 members. Coverage under this scheme is extended currently to more than 1,000 families.

Disaster insurance in Andhra Pradesh has been made affordable to low-income women with subsidies from two sources. Since 2000, the Indian regulatory authority has required insurers wishing to operate in India to service the low-income segment of society, and many insurers appear willing to incur a loss on their microinsurance business in order to access the broader market. Insurers such as the Oriental Insurance Company thus offer affordable contracts to low-income communities made possible by cross-subsidies from their other lines of business and wealthier clients. As a second source of subsidy, the UK-based donor NGO Oxfam paid 50% of the premium in the first year. Furthermore, Oxfam actively convinced the private insurer to offer very low cost insurance by training disaster management volunteers, who assist in providing insurance services such as helping communities in the claims process. Indian regulation and the success of NGO–insurer partnerships have motivated private insurance companies in India to actively pursue business with the poor (Krishna, 2005).

The Oxfam project, along with many other such programmes arising throughout Asia, Africa and Latin America (see Mechler et al., 2006), provides a second illustration of the types of disaster insurance schemes that could be supported by a climate insurance facility. Because of the covariant risk (raising the premiums due to costly back-up capital), disaster microinsurance is hardly affordable to low-income households or businesses without the kind of support recently institutionalized by Indian regulation and provided by donor organizations. This provides another example of how a climate insurance programme could support insurance in developing countries.

3.3. Assisting insurance schemes for private property in middle-income developing countries

Even in middle-income developing countries, such as Turkey and Mexico, many high-risk households and businesses cannot easily afford commercial insurance. Catastrophe insurance premiums fluctuate widely and are often substantially higher than the pure risk premium mainly because the insurer's cost of back-up capital is included in the costs. For example, in the Caribbean region, insurance premiums were estimated to represent about 1.5% of GDP during the period 1970–1999, while average losses per annum (insured and uninsured) accounted for only about 0.5% of GDP (Auffret, 2003). A formidable challenge facing insurance mechanisms in highly exposed developing countries is thus rendering them affordable to low-income clients.

3.3.1. Example: The Turkish Catastrophe Insurance Pool

The Turkish Catastrophe Insurance Pool (TCIP) launched in 2000 is the first of its kind to tackle the problem of insurance affordability in a middle-income developing country (see Gurenko, 2004). Istanbul faces a high probability of a severe earthquake (Parsons, 2004). In response to this risk, earthquake insurance policies are now obligatory for all property owners in Istanbul and other high-risk urban centres. Property owners pay a premium based in part on their risk-reduction measures, such as retrofitting their apartment buildings, to a privately administered public fund. The system does not apply to most of Turkey's very poor households by exempting property owners in rural areas. To reduce premiums and thus make the system affordable to urban dwellers, the World Bank absorbs a pre-specified part of the risk by providing a contingent loan facility with highly favourable conditions and contingent on the occurrence of a major disaster. In other words, if the fund cannot meet claims after a major earthquake, or if the earthquake is particularly catastrophic, the World Bank provides low-cost capital to the pool.

The TCIP would not have been possible without recent advances in catastrophe modelling. In the absence of large sets of historical data, advanced risk modelling simulation techniques have increased the confidence insurers place in risk estimates and have greatly enhanced the insurability of catastrophic risks (Kozlowski and Mathewson, 1997; Bier et al., 1999; Boyle, 2002; Clark, 2002). Although risk assessments can be very resource-intensive, by drawing attention to risk and prevention measures they can be useful beyond the pricing of insurance contracts. This is the case in Turkey, where local universities have worked together with government in assessing risks and drawing up a blueprint for prevention.

While the TCIP has received criticism about its imposition of mandatory policies, its somewhat weak link to risk reduction, and complications concerning illegal dwellings in Istanbul, this pioneering effort sets an important precedent as the first operational nation-wide disaster insurance system in a developing country. It has been made viable by an international financial institution providing technical support and absorbing a part of the risk. As such, the TCIP, like the micro-insurance schemes discussed above, provides another example of how a climate insurance facility can support developing-country insurance programmes. Although the TCIP addresses only earthquake risk, similar support could be extended to insurance systems that provide financial protection for floods, windstorms and other sudden-onset, climate-related disasters. This is the third example of how a climate insurance programme could be targeted to assist adaptation.

3.4. Assisting insurance mechanisms for public sector liabilities

Governments of highly exposed developing countries also need the assurance of sufficient funds to enable them to rebuild critical infrastructure and provide post-disaster relief. Without sufficient funds, the follow-on costs can be extensive. In the past, however, post-disaster sources of finance in developing countries have been woefully inadequate to assure timely relief and reconstruction. For example, 2 years after the 2001 earthquake in Gujarat, India, assistance from a government reserve fund and international sources had reached only 20% of original commitments (World Bank, 2003). International support for the India Ocean tsunami was exceptional, with estimates of about $7,000 per affected victim, which can be compared, for example, with the devastating floods affecting Bangladesh in 1998, where support was estimated at about $3 per affected victim (Tsunami Evaluation Coalition, 2006).

3.4.1. Example: Mexico's catastrophe bond

In Mexico, a taxpayer-supported national catastrophe fund (FONDEN) provides the government with needed funding for disaster relief. Since current and predicted reserves are considered insufficient for a major earthquake or other severe catastrophe, the Mexican authorities developed a mixed catastrophe bond and insurance risk-transfer strategy to protect FONDEN against catastrophic events, and in 2006 Mexico became the first sovereign country to issue a catastrophe bond (V. Cardenas, personal communication, 2006). A catastrophe bond is an instrument whereby the investor receives an above-market return when a specific catastrophe does not occur in a specified time (e.g. an earthquake of magnitude 7.5 or greater on the Richter scale in the vicinity of Mexico City over a 3-year period) but sacrifices interest or part of the principal following the event. The government's disaster risk is thus transferred to international financial markets that have many times the capacity of the reinsurance market. One major advantage of a catastrophe bond is that it is held by an independent authority and is not subject to credit risk. The payments go directly to the government, which in turn passes them on to FONDEN.

The development of Mexico's catastrophe bond was made feasible in the initial stages with technical assistance from the World Bank, but otherwise Mexico, as a middle-income developing country and member of the OECD, financed the bond out of its own means. This may not be possible for low-income countries, which presents another opportunity for assistance from a climate insurance programme (see Figure 2). This support can also take advantage of other financial instruments. For example, the World Bank has recently agreed to support the Colombian government's risk management plan with a contingent credit arrangement (World Bank, 2005b).

3.5. Summary

These examples demonstrate four types (among many) of interventions and support that could be offered by a climate insurance programme, either alone or in partnership with other donor organizations. These interventions include the provision of technical assistance, financial subsidies and reinsurance. It is worth emphasizing that this support, if financed through adaptation funds, would be based on recent and innovative precedents – on the part of the World Bank, the World Food Programme, Oxfam, and other donor and financial organizations. There are many possibilities for apportioning support; for instance, developing country governments and private initiatives might submit applications that are assessed according to specified criteria, such as their fairness, efficiency (in terms of risk reduction measures), practicality and governance (in terms of democratic procedure and stakeholder involvement in their design).

Interventions in the form of a climate insurance programme can render disaster insurance affordable in developing countries, and can also be tied to preventing risks. Furthermore, this assistance could increase the geographic spread and diversification of the risks (as shown in Figure 1), which is essential for assuring the robustness of insurance systems in terms of their capacity to absorb large or multiple shocks. This is particularly important for covariant losses that impose a substantial risk of insolvency on small insurance pools.

4. Thoughts on a second tier of support

As described above, a climate insurance programme for stochastic sudden- and slow-onset weather disasters, either through a climate insurance facility or other institutional arrangement, would not cover uninsurable risks or communities without access to insurance. In theory, all risks that can be reliably estimated and for which there is uncertainty with regard to their timing and consequences are insurable (Kunreuther, 1998); yet, in practice, insurance will not be offered for many types of catastrophic risks for reasons of both demand and supply. Risk perception and the lack of an insurance culture are two reasons, among others, that limit demand for even affordable insurance. Ambiguity in the risks and their estimates and problems of adverse selection (those most at risk tend to join the pool) are among the reasons for limited supply. Finally, insurance systems cannot be designed and implemented given insufficient data on the risks. Many highly exposed low-income governments and their citizens thus will not, at least not in the short term, be a part of an insurance programme. To include those who face uninsurable risks and those who are not able to insure, a second tier of support may well need to be considered (see Figure 1).

It should be emphasized that limiting support to sudden- and slow-onset weather risks may not fully meet the intent of the UNFCCC as expressed in Article 4.8, which specifically calls for actions to address the needs and concerns of small-island countries and countries with low-lying coastal areas exposed to gradual sea-level rise (United Nations, 1992). This proposal does not cover gradual (and, in terms of occurrence, predictable) risks; rather, it is meant to complement the AOSIS proposal, which is designed to compensate the victims of small-island states and low-lying developing countries for sea-level rise.

Support for persons or governments facing stochastic risks (as opposed to gradual impacts) for which no insurance is available could follow some of the principles set out by the AOSIS and Germanwatch proposals. Each advocates a fund fully financed by industrialized countries that provides post-event assistance contingent on pre-event risk or impact management efforts. There is one main drawback of these proposals: their wide scale of operations, including risk/property assessments and individual claim settling, limits their practicality. In contrast to the Germanwatch proposal, which would provide post-disaster funds by assessing claims, we suggest that an administrating authority should allocate swift post-event support based on pre-negotiated terms, but not based on the assessment of loss claims.

An appealing feature of both the AOSIS and Germanwatch proposed schemes is their attention to prevention by reducing vulnerability and the uncontrolled exposure of people and assets to hazards. Likewise, a second tier of support for our proposed strategy might consider imposing a criterion of eligibility requiring the qualifying country to demonstrate credible efforts at reducing, insuring and managing disaster and weather-related risks. These efforts would ideally be based on stakeholder-led efforts by the recipients to develop and implement risk management programmes that are designed within their own institutions and not imposed by an outside authority.

An important precedent for this second tier is the World Bank's planned Global Facility for Disaster Reduction and Recovery (GFDRR). While still under development, it is envisaged that one level of operation will provide technical assistance for assessing and mainstreaming disaster risk into development planning with country-wide risk management strategies; at another level, the GFDRR would be a stand-by facility to provide quick relief funding after an event. To some extent, this relief would be contingent on implemented risk management strategies (World Bank, 2006).

5. Challenges and opportunities

5.1. Challenges

Despite tangible benefits and novel prospects for supporting adaptation to the impacts of climate change through insurance-related instruments, there are many challenges and opportunities to negotiating financial resources for this purpose. A major stumbling block has been a call by the Organization of Petroleum Exporting Countries (OPEC) for parallel treatment (ECO, 2004). Just as AOSIS seeks financial assistance, insurance and the transfer of technologies under the UNFCCC to help small-island states and low-lying nations adapt to a changing climate, OPEC seeks compensation for lost revenues from the reduced use of fossil fuel. Although seemingly unrelated, negotiations on these two issues have long been intertwined and thus deadlocked. The linkage continues despite views that these two categories of impacts are different in kind, scope and temporal aspects, and different in the nature of the communities impacted (Barnett and Dessai, 2002). Vulnerable communities exposed to sea-level rise, threats to their water supplies or an increased intensity of hazards have played little role in creating these physical threats. In sharp contrast, the implementation of response measures can be expected to affect economies that have played a direct role in contributing to climate change (and that have benefited from this role), through fossil-fuel production or fossil-fuel consumption (for a detailed discussion, see Linnerooth-Bayer et al., 2006).

At the Seventh Conference of the Parties (COP-7) to the UNFCCC in Marrakech in 2001, it was agreed that predictable and adequate levels of funding shall be made available to developing country Parties to meet Convention commitments. It was also agreed that developed countries should provide resources through three newly created funds (Special Climate Change Fund (SCCF), Least Developed Country Fund, and Adaptation Fund), the Global Environment Facility, and bilateral and multilateral sources (UNFCCC, 2001). The creation of the SCCF was important in signalling a degree of political will to implement Article 4.8 and its related Kyoto Protocol provisions for the broad group of developing countries. The SCCF provides support for specified adaptation measures, including capacity building and institutional capacity for preventive measures, planning, preparedness and management of disasters related to climate change (UNFCCC, 2001).

The Marrakech funds are financed from diverse sources, including voluntary payments usually taken from Official Development Assistance (ODA) and the proceeds from a levy on the Clean Development Mechanism (CDM). Contributions to these funds have been made since Marrakech (Mace, 2005; Verheyen, 2005), but substantial funding has yet to be committed. The sentiment, especially on the part of developing countries, is that the COP has not created sufficient resources to address adaptation, despite the ample evidence of climate impacts in progress (Kartha et al., 2006). Alternative sources have also been proposed; for example, an international air travel adaptation levy (Müller and Hepburn, 2006).

5.2. Opportunities

It seems evident that any more ambitious form of support for insurance-related instruments in developing countries could benefit by partnering with financial institutions and donor organizations with similar aims. A consortium could link the proactive disaster-support agendas of multiple institutions, including international financial institutions (such as the World Bank, the InterAmerican Development Bank), bilateral donors (such as the UK Department for International Development (DFID) and the German Ministry for Economic Cooperation and Development (BMZ)), international organizations (such as the Organization for Economic Development (OECD), the United Nations Development Programme (UNDP) and the DG Development of the European Commission), reinsurers (such as Munich Re), and non-governmental organizations (such as Red Cross/Red Crescent and OXFAM). Coupling with other initiatives raises the question of the scope of climate adaptation funds committed to climate risk reduction. If funds for a climate insurance programme are pooled with support for seismic and other non-climate risks, this would have the advantage of increasing the global diversification and global benefits of the envisaged pool.

Two recent projects by the World Bank are especially promising as a potential link with the broad programme of support outlined in this proposal. As discussed above, the Global Fund for Disaster Reduction and Recovery (GFDRR) will provide technical assistance for mainstreaming disaster risk and serve as a stand-by facility to provide quick relief funding. A Global Insurance Index Facility (GIIF) sponsored by, among others, the European Commission, is in the planning stages. This facility, as envisaged, will provide backup capital for index-based insurance covering weather and disaster risks in developing countries to assure financial protection for small risk-transfer transactions. By constructing a diversified portfolio of developing country risks, the facility would leverage risk transfer and thus jump-start the development of risk transfer markets in countries with underdeveloped insurance markets (World Bank, 2005c). It is anticipated that other donor and financial institutions will join the GIIF initiative.

Another opportunity and challenge is to link insurance instruments with risk-reduction measures, and thus contribute directly to 'adaptation' (note that reducing long-term losses through a timely infusion of post-disaster capital also contributes to adaptation). Cleverly designed insurance systems can explicitly reward risk reduction behaviour with reduced premiums. With important exceptions, however, experience with incentive-compatible insurance is disappointing; yet, this record might be improved by setting risk reduction as a prerequisite for offering support. It should be emphasized that substituting pre-disaster support for post-disaster relief inevitably draws attention to the risks and opportunities for their reduction, and coupling insurance with risk management may have great potential for mitigating the human and economic losses from disasters (Linnerooth-Bayer et al., 2005).

6. Summary and issues

The importance of overcoming the impediments to progress on insurance-related actions within the UNFCCC and putting concrete proposals on the negotiating agenda has been emphasized by the first Executive Director of the UNFCCC, who considers implementation of Article 4.8 as 'one of the most critical aspects' of the climate-change negotiations (Capdevila, 2000, quoted in Barnett and Dessai, 2002). This article has made a tentative step in this direction by proposing the creation of a two-tiered climate insurance strategy, which would specialize in supporting developing country insurance-related initiatives and providing disaster relief for uninsured and uninsurable climate risks. The strategy could be implemented by a stand-alone mechanism, or it could leverage its support by partnering with other donor initiatives, such as the recent multi-donor initiative to create a Global Index Insurance Facility. Alternatively, it could be 'mainstreamed' into broader disaster risk management activities, perhaps by partnering with the World Bank's Global Facility for Disaster Reduction and Recovery.

Whatever institutional form the proposed climate insurance programme takes, its purpose would be to enable the establishment of public/private safety nets for stochastic shocks by making use of insurance instruments that are affordable in developing countries. In a second tier, the strategy could additionally, under specified conditions, provide disaster relief to those without an opportunity to join an insurance pool. A main advantage of this proposed two-tier strategy is that it takes advantage of the many opportunities provided by new and conventional insurance instruments, and combines individual, governmental and international responsibility. Moreover, its feasibility is demonstrated by donor-supported insurance programmes already under way. Finally, it offers opportunities for linking with other donor initiatives, providing incentives for loss reduction (adaptation) and targeting the most vulnerable.

The intent of this proposed strategy is not to provide a concrete proposal for negotiation, but rather to outline possibilities for further discussion. Many details and issues need careful consideration before a concrete proposal could be offered to the post-2012 negotiations. Most fundamentally, the institutional design of both tiers of this proposed strategy must be resolved: Should a climate insurance strategy take the form of an independent initiative, a partnered initiative, or simply a contribution to a disaster risk management fund, such as those recently set in train by the World Bank? There are also unresolved issues regarding the extent of support. If support is required to be commensurate with the incremental risk from climate change, it will be difficult to fashion a scientifically credible programme; alternatively, if support takes the form of a no-regrets strategy that integrates adaptation to climate change with adaptation to 'normal' climate variability, this two-tiered strategy, with its shared responsibility, offers many opportunities for differential levels of financial assistance. Finally, any discussion on a concrete proposal for allocating adaptation funds should necessarily be coupled with a discussion on the sources of this funding.

Acknowledgements

We are grateful for comments on earlier drafts of this article from Eugene Gurenko, M.J. Mace and Ian Burton. Of course, we take full responsibility for the content of this discussion.

References

Arnold, M., Kreimer, A., 2004. The socio-economic costs of disasters. In: A. Mathur, I. Burton, M. van Aalst (Eds), An Adaptation Mosaic: An Example of Emerging Bank Work in Climate Change Adaptation. Final Draft. World Bank Global Climate Change Team, The World Bank, Washington, DC, pp. 5–14.

Auffret, P., 2003. Catastrophe Insurance Market in the Caribbean Region: Market Failures and Recommendation for Public Sector Interventions. World Bank Policy Research Working Paper No. 2963. The World Bank, Washington, DC.

Bals, C., Warner, K., Butzengeiger, S., 2006. Insuring the uninsurable: design options for a climate change funding mechanism. Climate Policy 6, 637–647.

Barnett, J., Dessai, S., 2002. Articles 4.8 and 4.9 of the UNFCCC: adverse effects and the impacts of response measures. Climate Policy 2, 231–239.

Bier, V., Yacov, M., Haimes, Y., Lambert, J.H., Matalas, N.C., Zimmerman, R., 1999. A survey of approaches for assessing and managing the risk of extremes. Risk Analysis 19(1), 83–94.

Boyle, C., 2002. Catastrophe modeling: feeding the risk transfer food chain. Insurance Journal, 25 February [available at http://www.insurancejournal.com/magazines/west/2002/02/25/features/18828.htm].

Brown, W., Churchill, C.F., 2000. Insurance Provision in Low-Income Communities. Part II, Initial Lessons from Micro-insurance Experiments for the Poor. Microentreprise Best Practice, Bethesda, MD.

Brown, W., Green, C., Lindquist, G., 2000. A Cautionary Note for Microfinance Institutions and Donors Considering Developing Microinsurance Products. Microenterprise Best Practice, Bethesda, MD.

Clark, K.M., 2002. The use of computer modeling in estimating and managing future catastrophe losses. Geneva Papers on Risk and Insurance 27,181–195.

Cohen, M., Sebstad, J., 2003. Reducing Vulnerability: the Demand for Microinsurance. MicroSave-Africa, Nairobi, Kenya.

ECO, 2004. OPEC puts self interest ahead of G77 solidarity. ECOCOP 1004, 2.

Emanuel, K., 2005. Increasing destructiveness of tropical cyclones over the past 30 years. Nature 436, 686–688.

Government of Honduras, 2001. Poverty Reduction Strategy Paper. The World Bank, Washington, DC.

Gurenko, E., 2004. Introduction. In: E. Gurenko (Ed.), Catastrophe Risk and Reinsurance: A Country Risk Management Perspective. Risk Books, London, pp. xxi–xliii.

Hess, U., Syroka, J., 2005. Weather-based Insurance in Southern Africa: The Case of Malawi. Agriculture and Rural Development Discussion Paper 13. The World Bank, Washington, DC.

Intergovernmental Negotiating Committee [INC], 1991. A/AC.237/Misc.1/Add.3. UNFCCC, Second session, 19–28 June 1991. Geneva, Switzerland.

IPCC [Intergovernmental Panel on Climate Change], 2001. Climate Change 2001: The Scientific Basis. Contribution of Working Group 1 to the Third Assessment Report of the Intergovernmental Panel on Climate Change. Cambridge University Press, Cambridge, UK.

Kartha, S., Bhandari, P., van Schaik, L., Cornland, D., Kjellen, B., 2006. Adaptation as a Strategic Issue in the Climate Negotiations. European Climate Platform (ECP) Background Paper No. 4, Draft. Brussels, European Climate Platform (ECP).

Kozlowski, R.T., Mathewson, S., 1997. A primer on catastrophe modeling. Journal of Insurance Regulation 15(3), 322–341.

Krishna, H., 2005. Insurance for Vulnerability Reduction: Oxfam's Experience of Using Insurance as a Strategy for Disaster Risk Reduction in Coastal Andhra Pradesh-South of India. Background Paper for World Bank Institute on-line training course for India, Financial Strategies for Managing the Economic Impacts of Natural Disasters. The World Bank, Washington, DC.

Kunreuther, H., 1998. Introduction. In: H. Kunreuther, R.J. Roth, Sr (Eds), Paying the Price: The Status and Role of Insurance Against Natural Disasters in the United States. Joseph Henry Press, Washington, DC.

Linnerooth-Bayer, J., Mechler, R., 2003. Occasional Paper at UNDP COP 9 side event, Milan. 28 November 2003.
Linnerooth-Bayer, J., Mace, M.J., Verheyen, R., 2003. Insurance-Related Actions and Risk Assessment in the Context of the UNFCCC. Background Paper for UNFCCC Workshop on Insurance-related Actions and Risk Assessment in the Framework of the UNFCCC, 11–15 May 2003, Bonn.
Linnerooth-Bayer, J., Mechler, R., Pflug, G., 2005. Refocusing disaster aid. Science 309(5737), 1044–1046.
Linnerooth-Bayer, J., Mechler, R., Mace, M.J., forthcoming. Insurance and adaptation: opportunities and obstacles for the climate change negotiations. In: G. Sjostedt, F. Cede, G.O. Faure (Eds), Climate Change Negotiations. Earthscan, London.
Mace, M.J., 2005. Funding for adaptation to climate change: UNFCCC and GEF developments since COP-7. Review of European Community and International Environmental Law 14(3), 225–246.
Mechler, R., 2004. Natural Disaster Risk Management and Financing Disaster Losses in Developing Countries. Verlag für Versicherungswissenschaft, Karlsruhe, Germany.
Mechler, R., Linnerooth-Bayer, J., Peppiatt, D., 2006. Disaster Insurance for the Poor? A Review of Microinsurance for Natural Disaster Risks in Developing Countries. ProVention/IIASA, Geneva [available at http://www.proventionconsortium.org/themes/default/pdfs/Microinsurance_study_July06.pdf].
Müller, B., 2002. Equity in Climate Change: The Great Divide. Oxford Institute for Energy Studies, Oxford, UK.
Müller, B., Hepburn, C., 2006. An International Air Travel Adaptation Levy. Oxford Institute for Energy Studies, Oxford, UK.
Munich Re, 2005. Natural Disasters According to Country Income Groups 1980-2004. Munich Re NatCatSERVICE, Munich, Germany.
Opportunity International, 2005. Hard Times Don't Have to Equal Devastation [available at http://www.opportunity.org].
Parsons, T., 2004. Recalculated probability of $M \geq 7$ earthquakes beneath the Sea of Marmara, Turkey. Journal of Geophysical Research 109, B05304, doi:10.1029/2003JB002667.
Schönwiese, C.D., Grieser, J., Tromel, S., 2003. Secular change of extreme monthly precipitation in Europe. Theoretical and Applied Climatology 75, 245–250.
Siegel, P.B., 2005. Looking at Rural Risk Management using an Asset-based Approach. Paper prepared for the Commodity Risk Management Group, Agricultural and Rural Development Department, ESW. The World Bank, Washington, DC.
Tsunami Evaluation Coalition, 2006. The Tsunami Evaluation Coalition Synthesis Report [available at http://www.tsunami-evaluation.org/The+TEC+Synthesis+Report/].
UNFCCC, 2001. FCCC/CP/2001/13/Add.1, p. 43, Decision 7/CP.7 (Funding under the Convention).
United Nations, 1992. United Nations Framework Convention on Climate Change.
United Nations, 1997. Kyoto Protocol to the United Nations Framework Convention on Climate Change.
Verheyen, R., 2005. Climate Change Damage and International Law: Prevention Duties and State Responsibility. Martinus Nijhoff, Boston, MA.
World Bank, 2003. Financing Rapid Onset Natural Disaster Losses in India: A Risk Management Approach. Report No. 26844. The World Bank, Washington, DC.
World Bank, 2005a. Commodity and weather risk management programs to be expanded. In: World Bank News and Broadcast, 24 May 2005.
World Bank, 2005b. Colombia: Second Disaster Vulnerability Reduction Project. Report No. AB908. The World Bank, Washington, DC.
World Bank, 2005c. Managing Agricultural Production Risk: Innovations in Developing Countries. Report No. 32727-GLB. The World Bank, Agriculture and Rural Development Department (ARD),Washington, DC.
World Bank, 2006. Natural Disasters: Reducing Risk, Recovering Faster. The World Bank, Washington, DC.
World Food Programme, 2005. Pilot Development Project: Ethiopia Drought Insurance 10486.0. Projects for Executive Board approval, WFP/EB.2/2005/8-A, Rome.

Insuring the uninsurable: design options for a climate change funding mechanism

Christoph Bals[1]*, Koko Warner[2], Sonja Butzengeiger[3]

[1] *Germanwatch, Kaiserstrasse 201, Bonn, Germany*
[2] *United Nations University Institute for Environment and Human Security (UNU-EHS)*
[3] *Hamburg Institute of International Economics (HWWA), Hamburg, Germany*

Abstract

There is growing interest in the potential role that insurance-related instruments can play in the implementation of climate-change adaptation, particularly for the areas most affected and least able to absorb the negative effects of extreme weather events. Sufficient climate adaptation efforts will require funding at two or three orders of magnitude above the current levels. For rapid-onset climate events, current *ex-post* disaster finance does not offer strong incentives for risk reduction. This article suggests that insurance-related instruments can be a tool to help in adapting to and ameliorating the negative impacts of climate change for those countries likely to be most negatively affected by climate change. One possibility for an insurance-related mechanism would be a scheme that allows countries (or regions in large developing countries) most affected by climate change to purchase insurance-like coverage for defined climate-related risks. This article refers to such a scheme as the Climate Change Finance Mechanism (CCFM). The attempt to design and implement such an insurance-related mechanism requires careful consideration of several issues, including technical and political challenges. We outline a way to indemnify countries that are likely to suffer most from global climate change and consider what the key design elements would be.

Keywords: Climate change; Developing countries; Insurance; Scheme design; Adaptation finance; Risk reduction; Post-2012 negotiations

1. Introduction

Interest is growing regarding the potential role that insurance can play in the implementation of climate change adaptation, particularly for the areas most affected and least able to absorb negative effects. Sufficient climate adaptation efforts require mobilizing funding at the scale required, particularly given that the available funding is currently two or three orders of magnitude smaller than the levels needed (Loster, 2004). Insurance is considered as a possible way to increase the scale of funding (Indaba, 2006). Research further suggests that appropriately designed and implemented insurance

* Corresponding author. Tel.: +49-228-60-49217; fax: +49-228-60-49219
E-mail address: bals@germanwatch.org

mechanisms can bolster sustainable development and reduce poverty (Besley, 1995; Martin et al., 1999). Insurance can help anchor vital aspects of life quality – such as the ability to earn a livelihood and securing material assets (Dercon, 2004). Appropriately designed insurance can provide incentives for both public and private risk reduction, and contributes to a positive cycle of security and stability for those participating in the scheme(s) (Morduch, 1994; Holzmann and Jørgensen, 2000). Yet, so far, most people in developing countries have no access to affordable insurance coverage.

This article suggests that insurance-related instruments can be a tool to help adapt to and ameliorate the negative impacts of climate change for those countries likely to be most negatively affected by climate change. The article outlines one possible practical approach to indemnifying countries that are liable to suffer most from the global climate change. The idea of an international insurance pool covering climate change-related damage was raised for the first time by the Alliance of Small Island States (AOSIS) in 1991. In May 2003, critical issues and questions were raised at a United Nations Framework Convention on Climate Change (UNFCCC) workshop. Since then, however, the idea has not received much development. The proposed Climate Change Funding Mechanism (CCFM) aims to provide financial support for countries which suffer catastrophic economic damage due to climate change impacts and which cannot recover on their own due to a lack of financial resources. The challenge is to construct a financial compensation mechanism which would not only enable even the poorest countries to recover from damages caused by natural hazards, relying on the help of international community, but would also reduce their vulnerability to future natural disasters. The proposed scheme should also be politically acceptable for all parties involved and provide incentives for environmentally responsible behaviour by all countries, both in terms of adaptation/local prevention measures and in terms of emissions of greenhouse gases. It is not seen as a replacement for, but as one important tool in, an adaptation strategy.

2. Current disaster finance mechanisms

As a result of global temperature change, the frequency and intensity of extreme weather events is expected to increase, the impacts of which will differ regionally. Small island States and least developed countries (LDC) are likely to be among the countries most seriously impacted. According to the Third Assessment Report of Working Group 1 (IPCC, 2001), it is to be expected, with a high level of confidence (up to 99% certainty) that:

- both maximum and minimum temperatures will increase
- the number of hot days will increase
- there will be a change of regional precipitation patterns – leading to more intense rainfall events in some regions and more serious shortfalls or droughts in others
- tropical cyclone peak wind intensities will increase in some areas
- tropical cyclone mean and peak precipitation intensities will increase in some areas.

These prospects pose particular challenges for developing countries facing higher human and material losses (relative to GDP) from climate-related natural disasters (see Figure 1). There are few indications that private insurance can change this situation. In 1999, only 6% of economic losses in those 'emerging markets' were insured, while the share of insured losses in developed countries was 52% (Andersen et al., 2001, p. 7). Partly due to a lack of insurance and other risk transfer and finance mechanisms, weather-related losses are at present addressed through increased debt or international aid (Hoff et al., 2005).

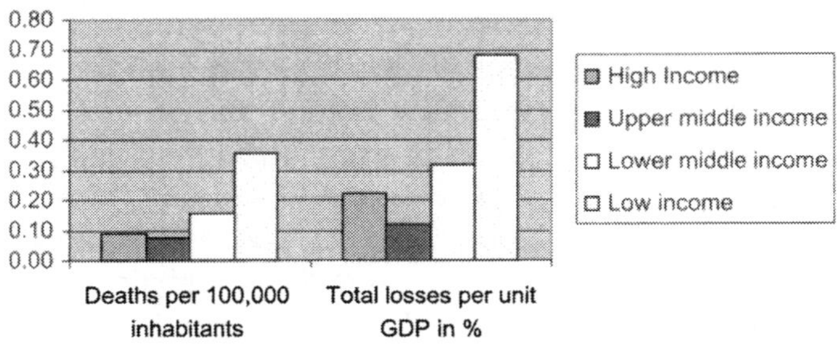

Figure 1. Relative affectedness (with respect to deaths and losses) caused by weather events in 2004, by country group based on income. The chart shows that the countries with lower per capita income are much more affected.

Developing countries currently finance losses with a variety of post-event mechanisms, primarily by allowing governments and victims to absorb the costs. Only a marginal use of insurance or other financial risk transfer mechanisms is utilized in the developing world today (Munich Re, 2005). Instead, donor aid has been a major source of risk financing for most disaster-prone developing countries. Over-reliance on this source of funding, however, also has major limitations, as outlined below. An insurance gap exists and could grow should the impacts of climate change become more apparent through extreme weather events such as floods, storms and droughts. Until recently, the amount of overall donor aid remained rather stable over time as a percentage of donor countries' GDP. Yet even with increases in ODA, economic losses caused by climate-related disasters over the past 20 years outstrip donor assistance by a factor of four to one (Warner, 2006).

3. Shortfalls of the current disaster risk financing models

Climate-related disasters are currently treated as exceptional, rare, unpredictable events. The perception persists that extreme weather events are unlikely to occur during any given political term, limiting the energy expended on fundamental rethinking of risk transfer for weather-related disasters (Warner et al., 2005). The factors that affect post-disaster assistance appear almost random:

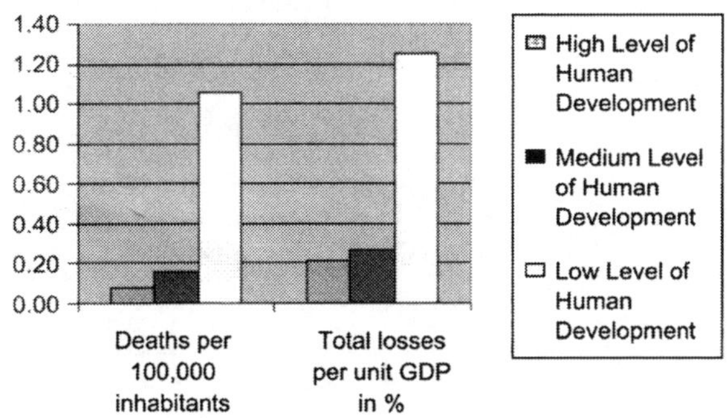

Figure 2. Relative figures for deaths and losses caused by weather events in 2004 for country groups defined on the basis of the Human Development Index (HDI). The chart shows that countries with a low level of development are much more affected.
Source: Anemüller et al. (2006).

typically a disaster must receive international media attention and tie in to a larger political issue. Additionally, the timing of an event and donor fatigue play a large role in determining the sum of resources received for disaster relief and recovery. The political priorities of donors also play a role in which countries receive disaster financing assistance today (Warner et al., 2005). For developing countries, in particular, a high hazard risk and low 'ability to pay' can cause economic growth to stagnate or decline, as illustrated in Figure 2. A 2002 study estimated that in the case of Honduras an additional $170 million of currently unbudgeted financing would be necessary in order to meet economic growth targets after a hurricane of Mitch magnitude (Freeman et al., 2002). In the absence of such finance, economic growth can stagnate and possibly decline in the long term due to the loss of capital stock (Figure 3).

Compared with the shortfalls of current disaster finance approaches, and given the expected rise in disaster losses related to climate change, the potential of alternative finance mechanisms should be explored. An appropriately designed and implemented insurance scheme for climate-related losses could be a strong improvement on the current approaches for several reasons:

- Insurance-related finance mechanisms require a conclusive dialogue about acceptable and unacceptable risks, values-at-risk, and specific risk reduction actions that a country or area would or should be willing to take in order to lower expected risks from climate change. This focus on risk transparency, communication about acceptable risks, and possible actions to limit exposure to risk is superior to the current *ex-post* system of funding natural disasters.
- Insurance-related finance mechanisms are not reliant on media attention to raise money. The money needed to repair and recover from disasters is released upon the occurrence of a predefined event or when insured losses reach a threshold specified in an insurance contract.
- Insurance-related finance mechanisms do not depend on the uncertainties of competition for donor funds when multiple events occur in different locations in the same donor year.
- Insurance-related finance mechanisms depend on defined parameters of loss rather than the 'fit' with political priorities of donor countries.

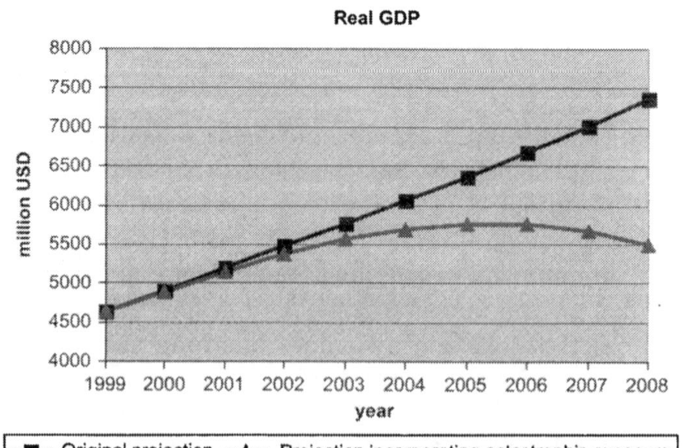

Figure 3. Impact of Mitch-magnitude hazard on Honduran economy in absence of sufficient financial resources for disaster relief and recovery. Source: Freeman et al. (2002).

One alternative for an insurance-related mechanism would be a scheme that allows countries (or regions in large developing countries) most affected by climate change to purchase insurance-like coverage for defined climate-related risks. This article refers to such a scheme as the Climate Change Finance Mechanism (CCFM). The attempt to design and implement such an insurance-related mechanism requires careful consideration of several issues, including technical and political challenges. The remainder of this article examines the general objectives, design aspects, and implementation issues of such a mechanism.

4. Objectives of the Climate Change Finance Mechanism (CCFM)

The objective of an international insurance-based mechanism is to provide financial support to developing countries affected by climate-related weather events. The objective of a Climate Change Finance Mechanism (CCFM) is the provision of financial support to sovereign governments for the rehabilitation of public infrastructure that has been damaged or destroyed by climate-related weather events (henceforth referred to as natural hazards). Under the proposed concept, countries with a high risk profile could *opt into* the CCFM system, which would enable them to cover infrastructure damage caused by natural hazards in situations where the country's (financial) capability is over-stressed, a capacity that could be defined upon entering into the CCFM system. At the same time, the system would provide incentives for increased adaptation and risk reduction measures, and mitigation of greenhouse gas (GHG) emissions.

5. Design of the CCFM: some key aspects

Various design options are available for CCFM as a public–private insurance-related mechanism. The following issues have been identified as key and most essential:

1. **Eligibility:** Which countries would be eligible to participate in CCFM and under what conditions?
2. **Coverage:** What will be the scope of CCFM's coverage?
3. **Structure:** What is the basic structure of CCFM?
4. **Incentives for risk reduction and adaptation:** How to provide incentives for adaptation/local risk reduction measures for developing countries?
5. **Participation:** How to create incentives for developed countries' participation?
6. **Operation:** Who might operate the CCFM?
7. **Financing:** How can the CCFM be financed?

The sections that follow briefly explore these issues.

5.1. Eligibility and conditions for inclusion

All regions with risk exposure to extreme climate-related weather events would be eligible to participate. For practical reasons, states are considered to be an appropriate participation entity, as they have the legal capability to negotiate international legal agreements. To reduce the overall amount of risk financing required for CCFM, (nearly) premium-free eligibility for financial protection from CCFM would be limited to less developed countries and small island states.

A country interested in joining CCFM can start 'opt-in negotiations' with the operator of the CCFM. While each individual country's case would be determined on the basis of these negotiations, it is important to establish in advance (i) what conditions a country must fulfil in order to become eligible for CCFM coverage; and (ii) under which circumstances the country would have the right to be indemnified by CCFM.

To create strong incentives for the country's own efforts towards local risk reduction, countries' eligibility for CCFM would be linked to a country's commitment to various risk reduction measures, including, but not limited to, hazard risk mapping, hazard zoning, and incentives for the population to relocate from disaster-prone areas. Implementation of these risk reduction measures can be made subject to regular international inspections to ensure compliance.

Once accepted into CCFM, countries would become eligible for financial protection on certain terms and conditions. Since CCFM's financial resources are likely to be limited, it would indemnify eligible members only in the case of extreme catastrophic events causing damages well in excess of international donor aid provided by the international community. This can be achieved by setting a pay-out trigger at the level of economic losses which could not be absorbed by the country without resorting to external assistance. This threshold is to be determined on a country-specific basis. Also, instead of an economic loss trigger, which may be subjective and difficult to verify in the aftermath of an event, one can think of determining pay-out triggers on the basis of physical characteristics of natural hazards (which are likely to trigger a certain level of economic loss according to hazard models) as long as they occur within predefined areas of the world. A combination of physical and/or economic indicators can be considered as well.

Pay-out thresholds (triggers) would be determined *ex-ante* based on objectively defined risk exposure characteristics of individual countries. These types of schemes are already being tested for weather-related risks, in both developed and developing countries (Loster, 2005). In India, one large NGO has established a rainfall index that allows subscribing farmers to receive payouts if they experience rain shortfalls during critical phases of the growing season (Warner et al., 2005). The main benefit of the trigger approach in developing countries is that proof of loss is not required, thus lowering transaction costs. The triggers can, however, be technically sophisticated, and the measurement stations must be tamper-proof in order to avoid cheating. Such schemes require longer-term observation and evaluation, but do provide examples of how triggers work for the type of scheme proposed. Such a country-specific approach to setting the terms of coverage would better meet the risk management needs of individual countries and making CCFM coverage more effective. The proposed approach would require detailed country risk assessments, which would lay the groundwork for designing the terms of CCFM coverage. (The National Adaptation Programmes of Action – NAPAs – established under the UNFCCC could constitute a basis for such risk assessments.) Such risk assessments will gather knowledge that can be used by governments to manage national catastrophe risk. Information gathered through country risk assessments conducted under CCFM can influence regional development plans with regard to placement of transport arteries and location of human settlements, and can affect the choice of construction technologies and building materials in the local construction industry. In the long run, this will help to reduce economic damage from future catastrophic events.

5.2. Scope of CCFM's coverage

One should differentiate between public assets – such as transport structures, water supply systems, waste water treatment, schools, public buildings, etc. – and private assets such as private houses and their contents, cars, boats, etc. Although it may be desirable to include both private and public assets under CCFM coverage, particularly in developing countries with an undeveloped insurance industry, this task may prove to be too complex to start off with. Instead, to reduce the complexity of CCFM, this article focuses on the coverage of public-sector assets. This, however, by no means implies that the task of covering private assets against the risk of global climate change is considered less important. Work on this aspect of coverage will be continued at a later stage.

5.3. Types of CCFM coverage

Although the main objective of the CCFM mechanism is to provide coverage for the most extreme catastrophic natural events, it can also be considered that it offers an additional (re)insurance coverage that would cover damages below the level of attachment of the basic coverage (e.g. below the damage threshold). Whereas, in the case of basic coverage, no monetary premiums would be paid to CCFM, extra coverage would be provided for an additional risk-based premium. Figure 4 illustrates the concept of the CCFM coverage.

5.4. How to provide incentives for local risk reduction measures in developing countries?

In the context of insurance, moral hazard refers to the increased probability of loss because of the insured's less risk-averse behaviour due to the availability of insurance coverage. In the case of countries, the reduced risk aversion might become a real problem, if not explicitly addressed by the appropriate design of terms and conditions of CCFM coverage. Such reduced risk aversion at the country level can manifest itself through construction of infrastructure in areas with a high risk potential for flooding (caused by excess precipitation or storm surge) or mudslides. Also, governments may decide not to invest in protection measures (e.g. the creation of dams or hazard zoning) if they know that the international community will compensate them for losses/damages anyway.

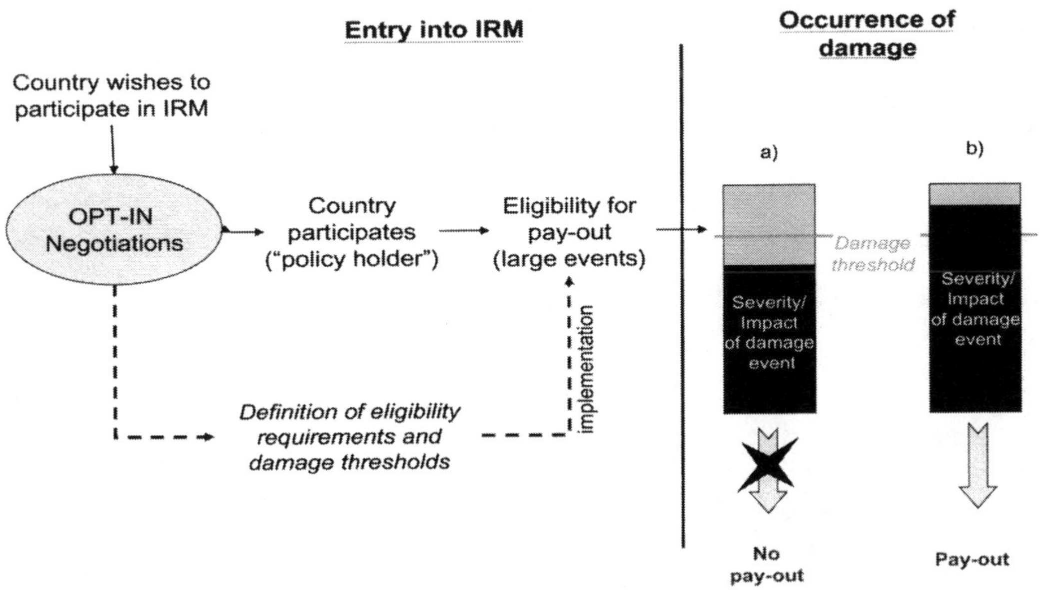

Figure 4. Concept of the Climate Change Funding Mechanism (CCFM).

In traditional insurance schemes, moral hazard is taken into account by insurance companies through prudent underwriting and by introducing deductibles and premium rates designed to discourage the insured from engaging in riskier behaviour. Such an approach may not be fully applicable to the proposed CCFM concept, as premium rates for the basic CCFM coverage are likely to be nil or very low. However, the proposed CCFM scheme would have access to other effective ways to address the moral hazard problem and to provide incentives for risk reduction efforts at the country level. These are as follows:

- The basic CCFM cover would cover damage from only very severe events, meaning that CCFM members would have to finance recovery from less severe events themselves. Consequently, the incentive to invest in risk reduction will be present.
- The CCFM would define minimum risk reduction measures to be undertaken by the country, where the annual cost to the country would be commensurate with the level of imputed risk-based premium for the type of coverage provided by CCFM to client members. The extent of adaptation measures necessary to qualify for CCFM basic cover would depend on the country's risk profile as well as its (financial) capability. By encouraging more risk-prone countries to invest relatively more in risk reduction projects or community adaptation (e.g. as a percentage of GDP), CCFM would be providing strong incentives for proactive risk reduction. Climate mitigation measures by developing countries through investments in emission reducing projects and technology could also count toward the country's in-kind premium contributions to the CCFM qualification requirements. One possible threshold could be linked to demonstrated investment in adaptation; however, especially for least developed countries, there must be flexibility to account for low-tech and low-investment adaptation measures. The cyclone preparedness programme in Bangladesh, for example, saved hundreds of thousands of lives by low-cost measures such as disseminating cyclone warning signals to local residents and assisting people in taking shelter (Munich Re, 2002). The investment in these programmes may be low in monetary terms, but their practical value is high for those who utilize them. An operational preference by international financial institutions towards large infrastructure projects is not necessarily the most effective solution. Meaningful indicators are needed to establish the degree to which investment in community-based preparedness programmes and other low-cost contributions meet the CCFM requirements.

5.5. Incentives for participation of developed countries

Below we provide a short overview of possible incentives for developed countries' participation in CCFM.

- By participating in the proposed CCFM, industrialized countries can show their commitment to the success of international climate negotiations.
- CCFM may be a viable way for industrialized countries to avoid or limit the scope of their potential legal liability for damages caused by global climate change in the future, as under the proposed system the calculation of financial contributions by countries could be based on their capability and common but differentiated responsibility, and not on the accountability for their emissions.

- By providing strong incentives for developing countries to implement risk mitigation measures, donor countries are likely to see a considerable reduction in the amount of post-disaster assistance requested annually from international donors by disaster-prone developing countries.
- A well functioning CCFM might constitute an alternative (free of moral hazard) model approach for the international community to provide disaster aid to disaster-prone developing countries after major disasters in the future – currently a big problem of post-disaster reconstruction lending and international donor assistance.
- CCFM may provide a real incentive for developing countries to reduce or contain emissions upon expiration of the Kyoto Protocol agreement in 2012, as investments in emission reduction projects could count towards the overall amount of in-kind country contributions required by the scheme.

5.6. Who might operate the CCFM?

CCFM would be operated by an independent operating entity that could be formed in the form of a public–private partnership (PPP) or by a private third party. One of the key advantages of a PPP is to blend core competences both from the private insurance industry and the public sector. While the key stakeholders, the organizational structure and work functions of the operating entity would be defined at a later stage, one of its core functions must be the development of terms and conditions for CCFM coverage, including coverage eligibility criteria. Determining the amount of in-kind contributions required under the CCFM programme (mitigation investments) and ensuring their adequate implementation by member countries would be yet another obvious task of the operating entity.

The opt-in CCFM scheme would also require local partners to plan and implement within the participating country. As this article focuses, as a preliminary step, only on the coverage of public-sector assets such as infrastructure (see Section on Scope of CCFM's coverage), the governments or regional authorities would be the expected partners.

5.7. Financing of the scheme

As in any insurance scheme, CCFM's policy holders would be generally expected to pay insurance premiums. The premium rates and coverage deductibles can be determined by CCFM on the basis of each country's risk exposure, including probability of occurrence and severity of potential damage, and scope of risk diversification for this risk globally. However, one very practical problem in this regard needs to be considered, namely a low or even non-existing ability to pay premiums by some countries. This is particularly true for least developed countries in Africa and Asia, which have very limited tax bases and which are often overburdened with debt. Consequently, it seems necessary that the coverage offered by CCFM is affordable for its poorest members. In addition, since the reduction of countries' physical vulnerabilities to natural hazards and their long-term capacity to adapt to climate change constitutes the main developmental goal of CCFM, it is essential that the scheme provides strong incentives for member countries to invest in risk reduction/climate adaptation projects. With these considerations in mind, we propose that, instead of asking countries to pay their premiums in cash, CCFM would be open to accepting premium contributions in-kind in the form of a country's investment in building their domestic climate adaptation strategy and mechanisms.

Therefore, financial support from the international community would be required to either subsidize countries' risk reduction projects and/or to provide risk capital for CCFM to finance its costs of reinsurance and consequently the costs of CCFM's coverage.

Below we consider different funding options for CCFM. We would like to make it clear from the start that only the 'basic CCFM' coverage would need to rely on additional financial support from the international community as, under the terms of basic coverage, member countries would pay little or no premium. Instead, country members would be required to provide in-kind contributions in the form of investments in domestic risk reduction projects or national risk-awareness and education campaigns.

The first possible source of CCFM financing is financial contributions by UNFCCC Parties. These contributions could be based on the criteria of capability (e.g. GDP/person) and/or on the aggregated amount of emitted CO_2 (per person) since a certain specified point in time (for instance, since 1990).

It should be noted, however, that from the political point of view the latter criteria is a very sensitive one. Many developed countries may be reluctant to support CCFM out of concern that such support would be perceived as a formal acknowledgement of their responsibility for climate change, which in the future could result in adverse legal consequences. But the use of the capability criteria would not produce very different results.

To finance their contributions to CCFM, developed countries may consider using a part of their general development assistance budgets – which is not likely to be appreciated by the developing countries. So as a more appropriate alternative, additional revenue-generating approaches may be explored by developed countries that, in turn, can set incentives for their own citizens to limit greenhouse gas emissions. These may include imposing climate fees on aviation or a general CO_2 fee. To finance CCFM, countries may also consider using at least a part of the income generated by auctioning emission rights within national emissions trading schemes. If the income generated by auctioning were used to set up regional or national adaptation funds, a window of these funds could be used for contributing to CCFM. One main benefit from these contributions to CCFM for its contributors would be the reduced vulnerability of developing countries to natural disasters in the future, and hence the ability to avoid diverting a growing percentage of development aid for emergency and disaster relief (1987/88 = 1.61% of ODA; 2003 = 8.51% – see also Hoeppe and Gurenko, 2006). By reducing the vulnerability of the world's poor to natural disasters, the CCFM contributors would also be making a meaningful contribution towards reaching the millennium development goals (MDG).

Among potential contributors to CCFM are international financial organizations such as the World Bank, Inter-American and Asian Development Banks, which have made the reduction of physical vulnerabilities in developing countries and their adaptation to climate change an integral part of their development agenda. Their financial support of CCFM could become an effective vehicle to promote hazard mitigation and risk awareness in their client countries. It should also be considered that the private (re-)insurance sector contributes to the CCFM.

6. Conclusions

The proposal for the global CCFM presented in this article has been intended as a contribution towards developing an interesting approach to funding natural disasters in disaster-prone developing countries. The proposed design features of CCFM aim to rectify numerous deficiencies of the existing model of disaster aid. One of the key problems with the current *ex-post* and *ad-hoc* form of international assistance is that it neither requires nor provides any incentives for effective risk reduction or climate adaptation measures on the part of aid-receiving countries. In the absence of effective risk reduction/adaptation measures, the increasing frequency and severity of natural disasters due to climate change is likely to extract even higher future tolls in terms of economic damages and lives lost in disaster-prone developing countries.

Building on the CCFM proposal presented in this article, the UNFCCC Parties can develop the legal and organizational framework for post-2012 negotiations with the view to extending the concept of global climate solidarity beyond the timeframe of the Kyoto Protocol. In their efforts, they are likely to be assisted by the global insurance and reinsurance industry, which has expressed a strong interest in establishing a global catastrophe insurance vehicle for developing countries.

References

ABI, 2005. The economic value of general insurance. Association of British Insurers [available at www.abi.org.uk/generalinsurance].

Andersen, T.J. et al., 2001. Managing Economic Exposures of Natural Disasters: Exploring Alternative Financial Risk Management Opportunities and Instruments. Natural Disasters Dialogue First Meeting, Washington, DC, November.

Anemüller, S., Monreal, S., Bals, C. [Germanwatch], 2006. Global Climate Risk Index 2006, Weather-Related Loss Events and their Impacts on Countries in 2004 and in a Long-term Comparison, p. 19.

Besley, T., 1995. Savings, credit, and insurance. In: J. Berhman and T.N. Srinivasan (Eds), Handbook of Development Economics, Vol. III. North Holland, Amsterdam.

Dercon, S. (Ed.), 2004. Insurance against Poverty. Oxford University Press, Oxford, UK.

Freeman, P.K., Martin, L., Mechler, R., Warner, K., 2002. Catastrophes and Development Integrating Natural Catastrophes into Development Planning. The World Bank, Disaster Management Facility, ProVention Consortium. Disaster Risk Management Working Paper Series No. 4. Washington, DC. October 2001.

Hoeppe, P., Gurenko, E., 2006. Scientific and economic rationales for innovative climate insurance solutions. Climate Policy 6, 607–620.

Hoff, H. Warner, K. Bouwer, L., 2005. The role of financial services in climate adaptation in developing countries. Deutsches Institut für Wirtschafts for schung. Special issue on the economic costs of climate change.

Holzmann, R., Jørgensen, S., 2000. Social Risk Management: A New Conceptual Framework for Social Protection, and Beyond. Social Protection Discussion Paper No. 0006. The World Bank, Washington, DC.

Indaba, 2006. Ministerial Indaba on Climate Action, 17–21 June 2006. South Africa.

IPCC, 2001. Climate Change 2001: The Scientific Basis, Contribution of Working Group I to the Third Assessment Report of the Intergovernmental Panel on Climate Change. Cambridge University Press, Cambridge, UK.

Loster, T., 2004. Rising costs of natural disasters and their impact on insurance. ProVention International Conference: Solidarity and Opportunity: The Potential of Insurance for Disaster Risk Management in Developing Countries. ProVention Consortium, Zurich.

Loster, T., 2005. Strategic management of climate change: options for the insurance industry. In: Weather Catastrophes and Climate Change. Munich Re, Munich.

Martin, I., Hulme, D., Rutherford, S., 1999. Financial Services for the Poor and Poorest: Deepening to Improve Provision, Finance and Development Research Program. Working Paper no. 9, IDPM, University of Manchester, October.

Morduch, J. 1994. Poverty and vulnerability. American Economic Review 84(2), 221–225.

Munich Re, 2002. Klima. Das Experiment mit dem Planeten Erde. Munich, p. 315.

Munich Re, 2004. Asset insurance premiums per capita per annum. Munich Re Economic Research, MRNatCatSERVICE. Munich.

Munich Re, 2005. Perspektiven: Ideen von Heute für die Welt von Morgen. Munich, 2005. p. 15.

Warner, K., Dannenmann, S., Ammann, W., 2005. Risk reduction (dis)incentives: findings of a survey in Latin America. Know Risk. ISDR. Tudor Rose, London.

Warner, K., Bouwer, L.M., Ammann, W., 2007. Financial services and disaster risk finance: examples from the community level. Environmental Hazards doi:10.1016/j.envhaz.2007.04.006.

The role of the private market in catastrophe insurance

Andrew Dlugolecki[1]*, Erik Hoekstra[2]

[1] *Climatic Research Unit, University of East Anglia, UK*
[2] *Munich Re, Munich, Germany*

Abstract

Globally, around 80% of disaster-related losses are uninsured. There are many reasons for this market failure: from the insurers' point of view these include high risk or small scale, absence of reliable risk data, and volatility in the event costs; from the at-risk population these include high prices, a misperception of the true risk, an expectation of government aid after disasters, and exclusion from financial services. We propose that a public–private partnership can resolve this. The public sector sets a framework to reduce the physical risks, provides cover for high-risk segments and regulates the market for other risks; the private sector provides consultancy and administrative services for all sectors and offers coverage for lower-risk segments. Competition reduces administrative costs and fraud. Cost-based pricing is an effective risk-management tool, and international (re)insurers can transfer knowledge and spread risks globally. A regional or global risk pool can reduce premium rates substantially by lowering the volatility of losses. Nevertheless the fundamental building block for catastrophe insurance is at the national level, since risks must be consistently estimated and administered locally.

Keywords: Public–private partnership; Catastrophe insurance; Market failure; Market mechanisms; Risk financing

1. Introduction

Risk financing can be provided from various sources, including the at-risk population, governments, donors, and, if conditions are right, the private insurance sector. Importantly, there are many essential non-risk-bearing functions such as claims handling that can be provided effectively by the private sector. A fruitful approach to explore is public–private partnership, where the public sector sets a rigorous framework to reduce the physical risks, provides cover for high levels of risk or segments with high administration costs, and sets the rules for a private market for other risks, while the private sector provides services and offers coverage for lower levels of risk and segments that are more easily accessible.

This article examines the current role of the private sector in catastrophe insurance, particularly for climate-related hazards, and analyses the key reasons for market failure. We then go on to propose an appropriate way for the public and private sectors to cooperate in providing insurance solutions for catastrophes. We conclude by considering how international insurance pools might work.

* Corresponding author. Tel.: +44-1738-626-351
E-mail address: andlug@hotmail.com

2. The current role of the private insurance market in catastrophe risk transfer

In order to understand how the private sector can participate in catastrophe insurance, one has to examine the key functions of an insurance market, understand the main private-sector actors in this market, its main customer segments, and the main risk exposures covered, as well as the insurance products offered.

2.1. Functions of an insurance market

In practice, insurance or risk transfer entails a number of processes, which can be 'unbundled' or combined in various ways. The private sector can participate in all, some, or none of these – examples can be found to illustrate almost every combination. At a strategic level, relating to the system design, a number of steps are involved (see Table 1).

Operationally, individual cases pass through different stages of underwriting (i.e. risk selection and assessment), pricing, setting contract conditions, loss adjustment, post-loss recovery and administration.

2.2. Private-sector actors

To undertake the key insurance functions, specialized 'actors' have evolved: insurers, reinsurers, risk modellers/actuaries, claims adjustors, agents/brokers and customer associations. Competitive forces drive the search for best practice and cost-efficiency, including fraud reduction.

The stake for each actor is not equal. For those concerned with administration or consultancy it is the operational cost, and opportunity cost of the capital, which are relatively modest compared with the potential revenue. For risk-bearers, however, the potential loss can be the whole of the capital provided, which is large in relation to the potential revenue. As the ratio of risk exposure to risk capital is potentially very large, due to the possibility of multiple, simultaneous losses, insurers and reinsurers may decide to participate only as risk administrators, rather than as risk takers. Catastrophe modellers will need to work with climate scientists to ensure proper risk assessment. Adjustors and brokers which are global operators can bring economies of scale, best practice and external resources in an emergency. Agents and customer associations may provide low-cost administrative solutions.

Table 1. Key design features of a climate insurance scheme

Design factor	Issues
Scope	Range of admissible risks
Financing	Source of capital and ongoing contributions (premiums)
Risk assessment	Data collection, development of risk models, pricing
Risk reduction/prevention	Regulation and enforcement of risk-reducing measures
Exposure control	Management of the aggregate risk to avoid insolvency
Product design	Stand-alone or integrated (bundled) cover, basis of indemnity, etc
Distribution	Networks to access customers and suppliers
Marketing	Consumer product education and incentives to join the scheme
Loss handling	Verification, fund transfer, dispute resolution, etc
Administration	Record-keeping, management of resources

A particular feature that has arisen in recent years is the involvement of non-insurance firms (e.g. banks and hedge funds) in this area through the supply of weather derivatives[1] and catastrophe bonds[2] to hedge against financial losses from abnormal weather. (Such contracts are not generally regulated as insurance products.)

2.3. Customer segments

Available insurance products vary for different classes of insured and by industry. The main insurance product lines are described below.

2.3.1. Agriculture and related activities

Agricultural production has numerous climatic risk exposures, e.g. drought, storm, pests, hail. Climate change alters these risks in two ways: through a slow-change process (mainly temperature and precipitation patterns) with positive and negative effects for agricultural productivity, and through an increasing number of extreme weather events. While the values at risk are higher in developed countries, the human and ecological consequences are greater in developing countries.

Traditionally, agricultural insurance systems have been supported by the government, because of the unusually high risk of moral hazard (when farmers exploit their insurance cover rather than making an effort to reduce their losses) and due to an anti-selection problem (when only farmers with substandard risks take out insurance). Also, the catastrophic loss potential often exceeds the capacity that the private insurance sector is willing to offer. However, while serving as financial safety nets for farmers, state-subsidized insurance schemes make it less likely that farmers will insure non-catastrophic losses privately, which limits the demand for private risk coverage.

2.3.2. Household catastrophe risk insurance

Climate models predict more intense hydrological cycles, and a steady rise in sea level. With an increasing concentration of people and values in coastal zones, mega-cities are becoming more vulnerable. This is particularly the case in developing countries, due to the poor control of land use and weak enforcement of building codes. Despite the growing risk exposures, insurance coverage against the risk of natural disasters in developing countries is almost non-existent. A part of the problem is that the poorer segments of population are typically bypassed by the insurance market. Microinsurance may be one way to overcome this. However, catastrophes can ruin such schemes, so it is important that they are considered in any system designed to deal with catastrophic risk.

2.3.3. Industrial/commercial insurance

The industries affected most by climate change are construction, energy, water, tourism and insurance, but also the food industry and apparel. Due to their own significant financial resources, large corporations often insure only a fraction of their risk exposures. In the USA, for instance, more than half of corporate risk is retained within 'captive' insurance companies controlled by the industrial corporations themselves. Insurance companies are now developing new products such as catastrophe bonds to effectively address the risk management needs of such large customers. Smaller enterprises use the private market extensively. Apart from very small businesses, the current private insurance market solutions are generally satisfactory for this sector but, even at this level, new products such as weather derivatives are appearing.

2.3.4. Public sector

The public sector faces a number of threats from climate change. Rising sea levels and river floods will mean an increasing cost for infrastructure. Extreme events such as storms are costly and divert resources from economic development. Natural resources, such as water and soil and the ecosystem, are also vulnerable to climate change. Global warming could also lead to increased heat mortality and infectious diseases, and hence increase health care costs for governments. Currently, in developed countries, insurance for these types of risk is rare. However, the ongoing privatization of utilities and transport leads to more demand for insurance solutions, as the privatized entities cannot carry uninsured risk exposures to natural disasters. In developing countries, as most of the infrastructure remains government-owned, most of that risk is currently retained by governments, which often resort to multilateral financial institutions and donors for disaster risk funding.

2.4. Main risk exposures

Typically, customers' insurance needs reflect their type of economic activity and vulnerability to abnormal weather or other extreme events. In the case of weather-related catastrophes, most of the insured loss comes from property damage, income interruption and liability for loss to others.[3] A key risk factor is the location of the assets, since natural hazards mainly affect particular geographical locations, e.g. coastal zones, flood plains, river valleys.

2.4.1. Property damage

The rapid growth in economic development and the increasing complexity of business processes (and hence their vulnerability to any disruption) create a rising demand for insurance cover against natural catastrophes. Three trends can be noticed: first, insurance companies are buying more protection, i.e. higher limits; second, companies are buying cover to protect against increased frequency of events; and third, they seek the inclusion of additional perils.

2.4.2. Income interruption

In the industrial and commercial sectors, abnormal weather or extreme events can cause loss of sales as a result of property or infrastructure damage. This is particularly true now, because businesses carry much less inventory, which makes them highly vulnerable even to the shortest interruptions in supply. In the case of individuals, a particular concern is loss of earnings in poor families if the wage-earners are incapacitated or killed.

2.4.3. Liability

Although natural disasters are, by definition, not caused by humans, human activity can be a major contributory factor, e.g. to erosion or flooding. One fear is that injured parties might seek compensation from businesses or States that have been perceived as being key contributors to climate change, and by inference extreme weather or climate-related events – for instance, entities that emit large amounts of greenhouse gases may be among the primary suspects. The same might be true for providers of financial services to such companies. Apart from the difficulties of attribution and causation, the potentially enormous costs would prevent any risk-bearing by the private insurance sector in respect of such a liability.

2.5. Design of insurance products

Traditionally, insurance policies require proof of loss for a claim to be paid, and the payment only restores the financial *status quo* for the policyholder. The contract depends upon payment of a premium by the policyholder to the insurer, who assesses the amount and sets relevant terms and conditions, including those related to the obligation of the insured with regard to risk management. In principle, the premiums from policyholders should be sufficient to cover claims, plus administration and a profit for the insurer.

The financial viability of conventional insurance products hinges upon the ability of insurers to apply differential pricing, different cover limits, deductibles and co-insurance to reduce moral hazard and anti-selection, which are costly and difficult to deal with. For that reason increasing attention is being paid to the use of parametric or 'trigger-based' products, where a specific loss to an insured party does not have to be proved for the payment to be made. Such index-based weather risk insurance contracts have emerged as an alternative to traditional crop insurance in developing countries. A more detailed discussion of these instruments is provided by Kelkar et al. (2006).

3. Insurance market failure

Globally, around 80% of disaster-related losses remain uninsured. There are many reasons. Commercial insurers are reluctant to provide coverage in the absence of reliable historical risk data. The cost of coverage can be disproportionately high due to market inefficiencies such as high administrative costs – up to 30% of the premium. Demand for insurance coverage by those at risk may be low due to ignorance of the true risks, or the expectation of government aid in the case of an event. Reinsurance costs are highly cyclical and volatile. Finally, the risk may be so high that it is, in principle, uninsurable. These problems often mean that pure private sector insurance solutions are not feasible.

To address these problems, governments sometimes intervene to ensure the economic viability of privately provided catastrophe insurance products either by offering premium subsidies or by establishing government-backed reinsurance facilities. In public–private insurance schemes, governments rarely act as direct insurers themselves. They prefer to rely on private insurers for distribution, while opting for other less direct forms of intervention. The most typical forms of government support for such schemes include legislation which makes catastrophe insurance compulsory, requirements for compulsory catastrophe reserves for insurers and reinsurers, or reinsurance backstop facilities to industry-run schemes, as in the cases of the Turkish Catastrophe Insurance Pool (TCIP) and the French '*catastrophes naturelles*' scheme. For poorer countries, public–private partnerships (PPP) in catastrophe insurance may also receive donor support in the form of free risk capital or subsidized insurance premiums.

In general, the key barriers to the ability of insurers to provide pure private catastrophe insurance solutions can be classified as supply-side and demand-side barriers.

3.1. Supply-side barriers

On the supply side, the main barriers to the participation of the private market in catastrophe insurance schemes lie in the area of *risk financing*. The main problems are as follows.

3.1.1. Volatility

Capital is the fundamental element for any insurance operation, as it ensures its ability to accept risks and pay claims. Capital mainly comes from private investors, who expect to receive a 10–20% risk-adjusted return. As insurance company's financial performance may be adversely affected by large claims from catastrophic events (which would prevent them from meeting their return targets), they make heavy use of reinsurance to stabilize their earnings. In the absence of affordable reinsurance capacity, insurers would be unwilling and unable to provide catastrophe insurance coverage. Alternatives such as equalization reserves[4] are, in principle, equivalent, but in fact the regulatory trend is towards abolishing these, because the modern accounting practice is to avoid financial transfers between years. Participation of the public sector in providing additional reinsurance capacity to the market is likely to reduce the price fluctuations of the reinsurance market, and hence would create a more stable and longer-term price stability on the reinsurance side.

3.1.2. Freedom to manage the underwriting process

A balance is needed between regulatory control of the market (to protect consumers), and flexibility in managing insurance operations in response to a changing risk landscape. To compete, companies need scope to design innovative products and to select clients according to the perceived risk. Geographical information systems (GIS) are increasingly used for locational underwriting of natural hazards. Overly rigid insurance regulations in local markets will deter private operators or result in less optimal insurance solutions.

3.1.3. Data availability on hazards and exposures

Poor data means that uncertainty is much greater and the private market will be less able to participate in risk-bearing. Geographical, economic and climate data tend to be poorer for developing countries, and access to such information is often prohibitively costly.

3.1.4. Risk prevention

In highly regulated markets, where insurers are limited in their ability to introduce appropriate risk-related discrimination among different risk classes of insured in terms of premium rates, deductibles and the scope of coverage, catastrophe insurance coverage may reduce consumers' risk awareness. It is therefore vital that public control of the risk management framework (land development, building design, construction standards, etc) is maintained and that regulators set a reasonable standard of care for policyholders in order to avoid such 'moral hazard'. The private sector can be a partner in this. In the UK, for instance, the insurance industry actively engages with policymakers on flood defence funding, land zoning and construction standards; while in the USA, insurers help to fund the technical training of publicly paid building inspectors. One way to overcome anti-selection is to make catastrophe cover compulsory or to bundle it with other essential finance products such as mortgage loans or fire insurance.

3.1.5. Administrative expenses

This is a major problem for policyholders with few assets. For example in the UK, although 80% of households have property insurance, this falls to under half for the poorest decile. The situation may be worsening with the decline of old distribution channels (local branch network, home service agents) and the spread of direct debit payments and the internet (Belson, 2006). For conventional products, a high minimum premium is necessary in order to reflect fixed costs per case, and also to avoid underinsurance. One solution is to offer insurance collection with rent, using the (often public-sector) landlord as a distribution channel. It is still necessary to demonstrate value for money to consumers, e.g. through subsidized community risk reduction, fast and effective claims settlement, and simple product structure (Hood et al., 2005; Perri 6 et al., 2005). In South Africa, other methods to reduce administrative expenses include payroll deduction of premiums, distribution through church networks, and simplified products and claim verification.

3.1.6. Synergy with other operations

Private market operators can gain significant economies of administration if they have a parallel operation that provides other products – e.g. fire or auto insurance – or can provide economies of scale from existing skill sets in other countries, such as risk modelling capability, policy administration systems, etc. This is particularly important for claim-handling, as capacity can be redirected from non-catastrophe work to assist in emergencies.

3.2. Demand-side barriers

There are various demand-side barriers as well. While some of them can be overcome by the private sector, over time, others may need public-sector intervention.

3.2.1. Perception of risk

Often consumers have low risk awareness of their risk exposures, particularly in the case of low frequency–high impact events. The private market can play a useful role in awareness-raising, since it has a profit motive to increase market penetration. As consumers are usually not willing to purchase disaster insurance, the introduction of compulsory disaster insurance by governments may be an important element in overcoming this problem.

3.2.2. Price

When premiums are high, consumers will not insure. This may be a signal from the private market that the risk is very high (unsustainable), or that there is great uncertainty, or that the scale of operations is too small, or that more risk management by at-risk parties is needed.

3.2.3. Availability and scale of publicly funded disaster relief

Often there is a public disaster relief system to cater for victims, e.g. emergency subsistence, soft loans. Unless it is carefully designed, this can undermine the viability of a private insurance market by reducing the demand for risk transfer.

3.2.4. Efficiency

The insurance process must be expedient – payment of claims must be achieved within acceptable timeframes or else consumers will not purchase the product. Here, private operators will seek to attract customers by being more efficient than their competitors.

3.2.5. Fairness

If consumers believe that they are paying more than their 'fair share' to the insurance fund, they will not insure willingly. The private market will seek to segment customers, thus eliminating cross-subsidies. However, this may be contrary to public policy in terms of solidarity.

4. Collaboration between public and private sectors

From the previous analysis, a public–private partnership (PPP) seems to be the appropriate model for insuring climate risk, because in developing countries public resources are limited. Table 2 outlines the respective roles of the public and private sectors.

Involvement of the private sector in disaster risk financing schemes has various advantages. For instance, risk-based pricing can be an effective risk-revealing mechanism, while the presence of international (re)insurers helps diversify the risk globally and transfer lessons internationally. Finally, with its emphasis on profit, the private sector tends to be more effective and innovative in its approaches to controlling administrative costs and fraud.

In addition, private insurers can offer cost-efficient products, marketing and distribution channels, as well as claims-handling systems. In light of its efficiencies, the private sector may be able to act as underwriting agent for the public sector or may perform a variety of services, even if it does not finance the risks itself. PPPs may also find it advantageous to cede at least a part of their catastrophe risk peak accumulations to the global reinsurance or capital markets.

5. International catastrophe insurance schemes

One way to deal with an annual, highly variable risk is to create an international pool where the premium rates can be substantially reduced by pooling the catastrophe risks of many countries in order to reduce the volatility of losses. To avoid anti-selection, i.e. losing the lowest-risk customers, the pool must strive to avoid cross-subsidies. The pool should also have a well-run professional claims settlement service in order to avoid a situation where claims rapidly overtake the premiums, causing the entity to collapse.

5.1. Local markets

Except in the case of governments directly insuring public assets with the pool, for an international pool to function there must be local insurance markets to carry out the necessary insurance functions

Table 2. Public and private sector roles in catastrophe insurance

Function	Public sector role	Private sector role
Risk assessment	Data collection, generic models	Risk modelling
Risk reduction measures	Regulation and enforcement	Product-based incentives
Product design	General regulation, consumer protection	All stages of product design
Risk financing (infrequent events)	Guarantee fund	Risk capital
Distribution/marketing	Consumer awareness, high-cost sectors	Multi-channel delivery
Loss handling	Minimal	Major role
Administration	Minimal	Major role

in compliance with local laws and regulations. Without the involvement of private insurers in the operation of a PPP, policies cannot be distributed, risks cannot be assessed, paperwork executed and funds collected or paid at the domain level.[5] The international pool would act as a 'reinsurer' that engages in only bulk or 'wholesale' transactions.

5.2. Regional pool

Where a supranational pool is to be used, the execution of the underlying functions of insurance is still required. To enable risk acceptance into the pool, the risks must be consistently underwritten and claims must be processed in a timely manner to ensure the pool's credibility with consumers. These insurance functions can be subcontracted by the pool to specialist insurance and risk management firms. Competition can be maintained by having more than one licensed firm, or by periodic tendering for services. The alternative is for the regional pool to carry out such work itself, which could be much costlier.

The risk portfolio of the pool can be built by offering several forms of coverage to different customer segments, as discussed above, but probably excluding industrial and commercial risks, which are adequately covered and where subsidies might infringe WTO regulations. Coverage forms are likely to focus on property loss and limited income protection.

There are two main options for routeing risks into the regional pool: either via a national pool, or via local private market insurers. The first method is preferable, since it reduces the number of transactions at the regional pool.

The regional pool is likely to act as a reinsurer of national pools for high layers of risk. The private sector could assist in the design of critical operational systems of the regional pool, such as systems for exposure management and wholesale transactions, and provide a range of other services. Alternatively, given that the regional pool is an efficient way to construct a portfolio of diversified risks, it may also be attractive to private-sector investors. Although, the pool itself is likely to be publicly funded, it might be able to 'lay off' some of the risk to the private reinsurance or capital markets through reinsurance contracts or alternative structures such as catastrophe bonds, which can be of interest to private investors.

5.3. Global catastrophe pool

If a global pool is to be established, it should function as a reinsurer of regional pools. The role of the private sector would be to provide 'wholesale' non-risk-financing services at the global level, and a wider range of insurance services together with risk-bearing at the regional and domain levels. Naturally, at the global level there will be a reduced choice of private-sector operators.

As the level of risk exposure in a global pool is higher in quantum, the direct participation of the private reinsurance market will be subject to actuarially sound pricing and the degree of correlation between the risk ceded from the pool and their own portfolios. If the example of the recently issued Mexican catastrophe bond is anything to go by, the global reinsurance market is likely to favour new risks from the emerging markets, as it would help to diversify their current risk concentrations in developed countries. Although the global capital market provides a much greater source of risk-bearing capacity, to date the insurance risk segment of the capital market remains small and the pricing of the risk is still, by and large, higher than in the traditional reinsurance market. At the global level, there are benefits for private-sector investors in terms of reduced volatility from the aggregation of diversified regional risks, which they can also obtain by participating in different regional pools. For some investors, the latter option may be preferable, as it provides for more control over the mix of risks that they are exposed to.

5.4. All-natural-hazard coverage

Another way to diversify risk is to extend the scope of coverage beyond weather-related disasters to include geohazards such as earthquakes and volcanoes; and including these perils can also help to achieve economies of scale. The private sector might also be more interested in providing some reinsurance capacity for a less correlated portfolio of risks. It is likely that an all-hazard system will be also more acceptable to consumers, as there will be a better take-up, due to an increased perception of risk.

6. Conclusions

The fundamental building block for catastrophe insurance is at the national level, since risks must be consistently estimated and dealt with in their everyday context to generate stakeholder confidence before aggregating them at the supranational level within regional or global markets. Design and preparation of new national insurance schemes can be greatly aided by the private sector, which can provide invaluable practical support on such issues as the collection of risk data, risk funding, underwriting, product design and administrative systems. At the operational stage, the private sector can also provide a wide range of support services and (possibly limited) risk financing.

Notes

1 A weather derivative is a contract that pays the buyer if a defined weather situation arises, e.g. a wet spell. No proof of loss is needed.

2 A catastrophe bond is a contract that pays investors regular interest. The buyer, usually an insurer or reinsurer, may cease the payments and even claim some of the capital, if a defined catastrophe happens, e.g. a category-5 hurricane. No proof of loss is needed.

3 Natural disasters also cause deaths, funeral expenses and increases in health care costs of course, but this article focuses on the classical property/casualty risks.

4 Money is put into and out of an equalization reserve when the actual claims are below or above expected levels in order to give a better measure of the long-term performance of a portfolio that is subject to erratic losses.

5 The domain is the level at which the legal structures governing the insurance policies are set; in most cases at country level, but other levels might apply, e.g. provinces in a federated country.

References

Belson, K., 2006. Rural areas left in slow lane of high-speed data highway. New York Times, 28 September.

Hood, J., Stein, W., McCann, C., 2005. Insurance with rent schemes: an empirical study of market provision and consumer demand. Geneva Papers on Risk and Insurance: Issues and Practice 30(2), 223–243.

Kelkar, U., James, C.R., Kumar, R., 2006. The Indian insurance industry and climate change: exposure, opportunities and strategies ahead. Climate Policy 6, 658–671.

Perri 6, Craig, J., Green, H., 2005. Widening the Safety Net: Learning the Lessons of Insurance With-Rent Schemes. Demos and Toynbee Hall. Commissioned by Royal and SunAlliance.

The Indian insurance industry and climate change: exposure, opportunities and strategies ahead

Ulka Kelkar[1*], Catherine Rose James[2], Ritu Kumar[3]

[1]*The Energy and Resources Institute, 4th Main, Domlur 2nd Stage, Bangalore 560071, India*
[2]*The Energy and Resources Institute, Habitat Place, Lodhi Road, New Delhi 110003, India*
[3]*The Energy and Resources Institute-UK, 27 Albert Grove, Wimbledon, London SW20 8PZ, UK*

Abstract

What is the preparedness of the Indian insurance industry to deal with the growing frequency and severity of natural disasters? We examine this question and argue that the continuation of present practices is not sufficient to address the challenges posed by climate change. The potential impact of climate change on the Indian economy can be severe, given the country's history of disaster losses, which is compounded by growth in population concentrations and burgeoning development in coastal and flood-prone areas. Targeted strategies are needed to deal with the rising costs of claims caused by climate change in a fledging Indian insurance market. The key challenges are to improve penetration of the available insurance products and to develop innovative delivery mechanisms to improve the access of the most vulnerable communities. Insurance is only a part of the solution, and must be combined with other measures that foster genuine preparedness and adaptation.

Keywords: Insurance; Climate change; India

1. Introduction

1.1. Climate change and India

India, with its large and growing population, densely populated and low-lying coastline, and an economy that has been closely tied to its natural resource base, is highly vulnerable to climate change. Two-thirds of the total sown area of the country is drought-prone, with monsoon rains showing high inter-annual, intra-seasonal and spatial variability. About 40 million hectares of land is liable to floods, with 8 million hectares and 30 million people affected each year on average (NCDM and NDMD, 1999). In the pre-monsoon and post-monsoon seasons, the coastline, particularly the east coast, is vulnerable to tropical cyclones. Over the period 1971–2000, India was among the top four countries in terms of number of people killed in natural disasters (Brooks and Adger, 2003).

* Corresponding author. Tel.: +91-80-25356590; fax: +91-80-25356589
E-mail address: ulkak@teri.res.in

India's first national communication on climate change impacts and measures, which was submitted to the United Nations Framework Convention on Climate Change (UNFCCC) in 2004, describes the potential impacts of climate change (Government of India, 2004). Climate projections indicate a marked increase in air temperature in the 21st century, which would become even more pronounced after the 2040s.[1] Models predict little change in total monsoon rainfall for India as a whole, but project an overall decrease in the number of rainfall days and an increase in rainfall intensity over a major part of the country. Preliminary assessments reveal a general reduction in the quantity of precipitation, and increase in severity of droughts and intensity of floods in various parts of India. Sea-level rise and higher storm surges may adversely affect coastal ecosystems and structures, leading to losses of settlements, property, recreation beaches and tourism infrastructure. Extreme events such as droughts, floods and cyclones may become more frequent, leading to widespread damage to life, property, crops and livelihoods.

1.2. Natural disasters in India

The potential impact of climate change on the Indian economy can be severe, given the country's history of disaster losses (Table 1), which is compounded by the growth in population concentrations and burgeoning development in coastal and flood-prone areas.

The World Bank estimates that direct annual economic losses from natural disasters, most of which are climate-related, amount to 2% of India's GDP (Gurenko and Lester, 2003). It also observed a rising volatility in reported monetary losses from natural disasters (Figure 1).

These findings were driven home by the July 2005 floods in Mumbai, India's commercial capital, caused by a record level of precipitation within 24 hours unprecedented in the country's recorded weather history. The floods resulted in the record economic loss of $5,000 million of which insured loss was about $770 million or about 15% (Munich Re, 2006). While this unusually high percentage of insured losses compares favourably with the small insurance loss observed in the Gujarat earthquake, we must point out that in the case of the recent Mumbai floods most of the insurance payments went to large businesses rather than home-owners, as the level of insurance penetration for the industry is many times that of residential uptake.

In the case of natural disasters in India, typically attention has been focused on relief and reconstruction in the aftermath of a disaster rather than on prevention and preparedness. The affected state government manages relief work and reconstruction efforts with financial support from the central government through the Calamity Relief Fund (CRF), with additional assistance from the National Calamity Contingency Fund (NCCF) in the case of severe calamities. However, financial discipline is an issue, which is obvious from frequently inflated demands for post-disaster financial

Table 1. Disaster history by major climate hazard in India during 1996–2001

Hazard	Reported events	Reported deaths (000)	People affected (000)	Reported losses (USD million)
Windstorm	15	14.6	252,13.7	5,619
Flood	29	8.9	150,980.3	2,928
Drought	4		90,000	588
Other	24	5.9	356.9	

Source: Gurenko and Lester (2003).

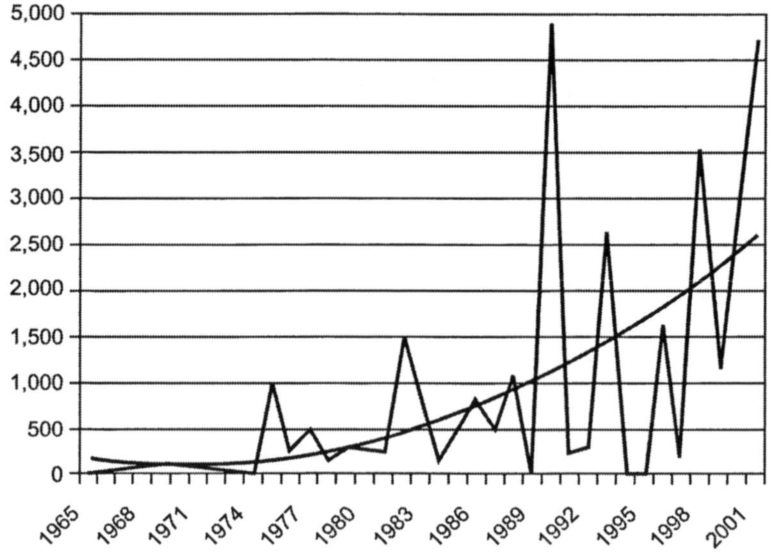

Figure 1. Reported direct losses from natural disasters in India during 1965–2001 in nominal US$ million at then applying exchange rates. Direct losses refer to stock losses including destruction of public and private economic infrastructure, productive capital, and housing. Source: Gurenko and Lester (2003), based on data from OFDA/CRED (2003).

assistance from the state governments, late contributions from states to the CRF, and delays in transferring funds to the district level. Declaring districts drought-prone is a politically sensitive issue because of the implications for financial assistance, and the possibility of receiving various waivers and concessions (including food-for-work schemes and rescheduling of short-term agricultural loans). In addition, the World Bank has pointed out that the current approach lacks institutional incentives and underplays the role of risk financing through *ex-ante* mechanisms (such as catastrophe reinsurance and contingent credit facilities) that could provide financial liquidity in the aftermath of natural disasters and kick-start economic recovery (Gurenko and Lester, 2003).

2. Role of insurance in disaster loss financing in India

Despite the fact that India is the second most disaster-prone country in terms of the frequency of these events, the level of non-life insurance penetration is abysmally low, under 1%, even when compared to countries with a similar level of GDP. While insurance plays a major role in compensation for disaster-related losses in developed countries, accounting on average for over 40% of economic losses, it is only a minor source of disaster risk financing in India. Government by and large remains the main financier of disaster relief, rescue, rehabilitation and reconstruction efforts. However, in the aftermath of the Gujarat earthquake in 2001,[2] the Indian psyche has shown a slight but discernible shift from a reactive to a more proactive management of disasters, in which insurance has an important role to play, as reflected in government policy and legislation such as that aimed at revisions in building codes for risk mitigation (MoHA, 2005). In the context of the increasing volatility of weather patterns due to climate change, insurance can smooth out the adverse financial consequences on households, businesses and the government. Insurance can also prove to be an important adaptation tool, which, over time, can help reduce India's vulnerability to natural disasters.

2.1. Present status of the Indian insurance industry

Although the Indian insurance industry remains underdeveloped, as can be seen from the low levels of insurance penetration and insurance consumption per capita, it has shown signs of improvement since its partial deregulation in 1999 (Infrastructure Regulatory and Development Authority Act, 1999). Besides lifting the ban on private players and opening the industry to foreign players in a limited manner, the Act also established the IRDA to oversee and regulate the insurance industry and named the General Insurance Company (GIC) as the national reinsurer to which all the country's direct insurers must cede 20% of their business. In the aftermath of these partial liberalization measures, insurance penetration has been on the rise, although it is still extremely low. In 2003, total insurance premium was US$9.9 billion, or 2.3% of GDP or US$16.40 per capita, and holds a very large growth potential (Table 2). Yet, the market for insurance, even after 4 years of liberalization, comprises only 29 companies, including one reinsurance company, as compared with 61 in China at the end of 2003. This may be taken as an indication of the inadequate competition in the Indian insurance sector in the light of the huge market potential. The current 26% restriction on foreign ownership in local insurance companies has limited the involvement of global insurance companies in the Indian insurance industry. It is hoped that recent talks on increasing the cap on FDI to 49% will attract more global players to invest in India and also encourage the growth of existing players by enhancing product innovation and service levels.

2.2. Coverage of natural perils in life, property and vehicle insurance

Except for personal accident policies, which cover accidents triggered by natural hazards, traditional insurance products do not cover property damages/loss or injury/death arising out of natural perils. Although both motor and property insurance provide the insured with an option of adding a natural-hazards coverage (which includes storm, tempest, flood and inundation) for an extra premium, most insurance buyers decline this extra coverage in an effort to reduce the premium.

2.3. Reasons for low insurance penetration

An underdeveloped insurance market makes it extremely difficult to diversify the risks arising out of large disasters, such as a tsunami, or a series of small disasters. The low insurance penetration in the country can be traced to a number of factors.

Table 2. Indian insurance industry: trends

	2000	2003
Insurance penetration (% households)	2.32%	2.88 %
Insurance consumption (USD per capita)	9.90	16.40
World rankings in premium volumes	23	19
Share in world market	0.25%	0.29%

Source: Rao (2004).

On the demand side, the biggest hurdles are the very low income of the population and the common perception of insurance products. Insurance in India has been traditionally sold more as a savings instrument rather than a risk protection vehicle, with tax incentives offered on a life policy to the individual assessee. Coupled with this is a low awareness among the public about insurance products in general. As a result, personal risk management is usually reactive and, in case of catastrophes, episodic. The experience of major insurance companies shows that following a major catastrophe, there is a rush for insurance cover, particularly for life and assets such as property, motors, etc. But this interest is short-lived and in the majority of cases these policies are not renewed (Banerjee, 2001).[3]

Another major factor is that large sections of the Indian economy operate outside the formal economy – not just small businesses, but also housing (slums). All insurers are required to provide some coverage for the rural sector. In addition, each company is obligated to service the social sector, which includes the unorganized sector, informal sector and economically vulnerable or backward classes in rural and urban areas. However, a few large group policies to agribusiness account for a large proportion of the insurance sales under this requirement (Sinha, 2005).

On the supply side, easy access to insurance products is still an issue. Distribution is critical in enhancing penetration and can have a huge impact on profitability, product design and, most importantly, the cost of insurance (premium levels). However, until recently, distribution of insurance was largely carried out directly by the insurance companies' sales force or through tied agents, which limited the industry's ability to reach a wider customer base. This has been particularly true in rural areas, which according to the latest census account for 72% of the total population. In contrast, in urban areas, the high population density helps the companies to reach customers with relatively less effort and expense through a more compact network. While the recent efforts to look at innovative ways of reaching the rural population through e-Choupals (village Internet kiosks), self-help groups (SHGs) and bancassurance models have been successful, these are yet to be scaled up, bearing in mind the large percentage of the rural population with no access to insurance at all.

Although the poor in rural areas have higher disaster risk exposure and also suffer more than their urban counterparts from natural disasters, their awareness of formal risk mitigation measures remains very low. Coupled with their limited purchasing power and volatility of income, this translates into a low expressed demand for insurance. In fact, scalability of successful insurance initiatives for the poor poses the most formidable challenge of all, which also necessitates the importance of designing insurance products in a custom-tailored fashion to meet the needs of the rural poor. The situation is further exacerbated as incentives to purchase insurance are further dampened as governments and other donor agencies often compensate losses to the poor on account of disasters. Not surprisingly, the nascent urban-centric Indian insurers have yet to make a dent on the rural populace in spite of the fact that improved access to insurance coverage for the poor has been integrated into insurance law (IRDA Subregulation 3, 2002).

Unfortunately, in the case of natural disasters affecting an entire community or area, traditional risk-sharing strategies (such as borrowing from relatives or moneylenders, selling assets, reducing consumption or inputs in farming) break down (Hess et al., 2002; Lilleor et al., 2005). In an effort to reduce their overall risk exposure to natural disasters, rural households often try to diversify the sources of their income by taking employment outside agriculture in industry or services. In the absence of formal insurance indemnification (due to the lack of access to suitable and affordable insurance products) in the aftermath of natural disasters, the rural poor are forced to turn to moneylenders or sell their productive assets, which frequently undermines the prospect of recovering their livelihoods. Another strategy is to send family members to work elsewhere and remit payments (Conroy et al., 2001; Deshingkar and Start, 2003; Rosenzweig and Stark, 1989).[4]

3. Current climate risk insurance approaches in India

3.1. Public sector approaches to catastrophe risk management

Traditionally most of losses caused by natural disasters have been addressed through government assistance or informal risk-sharing at the community level. Due to the very limited insurance penetration, the insurance industry in India has played a very small role in dealing with the impacts of either climate variability or extreme events such as droughts, floods and cyclones.

3.2. Government crop insurance

The Ministry of Agriculture has a National Agricultural Insurance Scheme, administered through the Agriculture Insurance Company (AIC), which covers all farmers irrespective of their size of landholding. Introduced in 1999–2000, the scheme operates on the basis of an 'area yield approach' by indemnifying owners of crops in areas which have suffered from calamities such as floods, hailstorms, landslides and cyclones (as evident by their lower yields compared with historic averages). Although government crop insurance is heavily subsidized, it has very low coverage. It fails to provide the right incentives to farmers as crop yields are insured irrespective of their risk mitigation efforts. The programme also has a major drawback by design – farmers who have suffered losses as a result of lack of rainfall in a particular part of a district may not be eligible to benefit from crop insurance unless the entire district is declared drought-affected. Finally, the programme has high administrative costs and is notorious for long delays in paying claims. The main reasons are (1) a highly complex institutional structure, involving several national and state government departments and public sector agencies, which leads to delayed flows of information and funds, and (2) the time taken to carry out extensive crop-cutting experiments to determine yields for different areas (for instance, about 40,000 such experiments are conducted annually in the state of Karnataka) (Kalavakonda and Mahul, 2005).

3.3. Alternative risk-management products: weather-indexed contracts

The recent partial liberalization of the Indian insurance market has cleared the way for product innovation in both life and non-life segments of the market. In case of non-life, various innovative products, including those aimed at dealing with the risk of climate variability and mostly oriented towards farmers, have been introduced. Some of these insurance products which are pertinent to the management of climate change in India are reviewed below.

Internationally index-based weather risk insurance contracts have emerged as an alternative to traditional crop insurance. These are linked to the underlying weather risk defined as an index based on historical data (e.g. for rainfall, temperature, snow, etc) rather than the extent of loss (e.g. crop yield loss). As the index is objectively measured and is the same for all farmers, the problem of moral hazard is minimized, the need to draw up and monitor individual contracts is avoided, and the administration costs are reduced. Weather-indexed insurance can help farmers avoid major downfalls in their overall income due to adverse weather-related events. This improves their risk profile and enhances access to bank credit, and hence reduces their overall vulnerability to climate variability. However, one of the inherent disadvantages of weather-derivatives is that, because of the way the index is defined, there may be a mismatch between payoffs and actual farm losses (CRM, 2005; Hess, 2003b); the problem being known as a basis risk. The strengths and weaknesses of traditional and weather-indexed agricultural insurance are summarized in Table 3.

Table 3. Comparison of traditional and weather-indexed crop insurance

	Traditional crop insurance	Weather-indexed crop insurance
Design aspects	• Can protect farmers against adverse weather conditions • Fails to provide the right incentives as crops yields are insured irrespective of risk mitigation efforts	• Protects the farmer's overall income rather than the yield of a specific crop • Moral hazard and claims manipulation eliminated by use of objectively calculated index
Implementation aspects	• High administrative costs and long delays in claim settlements • Failed in India in terms of coverage of farmers and financial sustainability	• Quick payouts triggered by independently monitored weather indices can improve recovery times and enhance coping capacity • Implemented on pilot basis by trying out different types of delivery models • feedback from engagement of local institutions such as MFIs

As discussed earlier, insurance distribution channels have a significant impact on determining access to insurance products. Various types of distribution models are currently being considered. These include direct marketing of weather-derivatives to farmers, or through banks, cooperatives, MFIs or input suppliers. Unlike traditional crop insurance, where claim settlement may take up to a year, quick payouts in private weather insurance contracts can improve recovery times and thus enhance farmers' coping capacity. Various pilot schemes and new initiatives that are currently on the way in India are discussed next.

3.3.1. ICICI Lombard pilot scheme for groundnut in Andhra Pradesh

ICICI Lombard General Insurance Company, with support from the World Bank and International Finance Corporation, conceptualized and launched a pilot rainfall insurance scheme in Mahabubnagar, Andhra Pradesh, in July 2003. The district had previously experienced three consecutive droughts. The scheme was implemented through the KBS (Krishi Bima Samruddhi) local area bank of BASIX, one of India's largest microfinance institutions. KBS Bank bought bulk insurance policies from ICICI Lombard and sold around 250 individual policies to groundnut and castor farmers. The index capped rainfall level per subperiod at 200 mm, and weighted critical periods for plant growth more heavily than others. The premium rates are defined in Table 4. KSB decided that only borrowing farmers could buy weather insurance policies. Eventually KSB planned to lower the interest rate on its loans to the insured farmers due to the reduced default risk (Hess, 2003a).

ICICI Lombard also launched a pilot scheme for insurance against excess rainfall for rice farmers in Aligarh, Uttar Pradesh.

Table 4. Premium rates (in Rs) for groundnut farmers in Mahabubnagar pilot scheme

Landholding size	Premium rate	Maximum claim
Small (less than 2 acres)	400	14,000
Medium (2–5 acres)	600	20,000
Large (more than 5 acres)	900	30,000

3.3.2. ICICI Lombard loan portfolio insurance

ICICI Lombard, assisted by the World Bank, developed weather-based loan portfolio insurance in India. In July 2004, the first weather insurance policy was offered to Bhartiya Samruddhi Finance Ltd (BSFL), the non-banking finance arm of BASIX group. ICICI Lombard would compensate BSFL for deviations in rainfall below the threshold level, which is fixed as a percentage of the average rainfall in the area. This is the first instance of an agricultural finance institution transferring the systematic risk of its crop lending portfolio to the insurance weather risk market.

3.3.3. ICICI Lombard weather insurance for salt production

In March 2005, the first non-agriculture-related weather insurance cover was provided by ICICI Lombard to Gujarat Heavy Chemicals for a sum insured of Rs15 million. In the event of loss of salt production due to unseasonal rainfall at its Nagapattinam salt production fields in Tamil Nadu, ICICI Lombard would pay Gujarat Heavy Chemicals an insured compensation of Rs0.6 million, Rs1 million or Rs2.5 million per non-production day, depending on the frequency, duration and severity of the interruption. The premium rates were arrived at by ascertaining various components such as expected loss, volatility of loss, capital costs and the company's profit and overhead costs.

3.3.4. KBS pilot scheme for soya farmers in Ujjain

BASIX/KBS also designed policies for soya farmers in Ujjain, Madhya Pradesh, which linked insurance to bank loans. In normal years, soya farmers taking crop loans of Rs2,000 with embedded weather insurance would be charged an interest rate of 20.5% instead of 17.5%. However, when cumulative weighted rainfall during the critical growing periods fell below 80% of the mean, farmers would receive interest payment relief of Rs10 per every millimetre of rainfall index deficit. Thus farmers pay a higher interest rate in normal years as the weather insurance premium, but receive much-needed relief in drought years (Hess, 2003b).

3.3.5. Rajasthan government insurance for the orange crop

ICICI Lombard General Insurance Company entered into a collaboration with the government of Rajasthan in June 2004 to provide rainfall-indexed insurance for orange growers in Jhalawar district and adjoining areas. The scheme covered two types of perils, with premiums defined in Table 5.

The insurance policy was made available through branches of the Land Development Bank and Jhalawar Cooperative Bank, rural branches of commercial banks in Jhalawar, Jan Mitra kiosks,

Table 5. Premium (in Rs) for units of sum insured of Rs5,000

Type of peril	Premium at 50% discount to small and marginal farmers
1. lack of effective shower to initiate flowering	415
2. dry spell during flowering	315

and direct sales agents of ICICI Lombard. Claim settlement was to be completed within 30 days of the expiry of the cover period. The Government of Rajasthan has also asked interested insurance companies to send proposals for similar schemes for oranges, cumin, coriander and other crops (Government of Rajasthan, 2005).

3.3.6. IFFCO-TOKIO monsoon insurance
In July 2004, IFFCO-TOKIO General Insurance Company Ltd (the Indian insurance arm of the Millea Group) announced plans to launch an insurance scheme for deficit rainfall during the monsoon months. Named Barish Bima Yojana, the scheme was targeted at four states – Gujarat, Maharashtra, Andhra Pradesh and Karnataka.

4. Findings from recent experience and lessons learned

4.1. Results update

The first experiences with selling weather-derivatives to farmers have been somewhat mixed. According to ICICI Lombard, while the insurance of the rice crop against excess rainfall in Aligarh and the insurance of oranges in Rajasthan were profitable, this was not the case with the insurance of groundnut crops against rain shortfall in Mehboobnagar, implemented by BASIX. The BASIX website provides data about its claims service performance (see Table 6).

The basic problem is that of high administration costs incurred while selling the insurance to individual farmers. ICICI Lombard finds it uneconomical to seek out each farmer, but would

Table 6. Rainfall insurance business and claims performance

<table>
<tr><th></th><th>April 2004–December 2004</th><th>April 2003–March 2004</th></tr>
<tr><td>Customers</td><td>427</td><td>300</td></tr>
<tr><td>Acres</td><td>670</td><td>450</td></tr>
<tr><td>Premium (Rs)</td><td>150,000</td><td>100,000</td></tr>
<tr><td>Average sum insured (Rs)</td><td colspan="2">6,000</td></tr>
<tr><td>Claims reported</td><td colspan="2">185</td></tr>
<tr><td>Claims settled</td><td colspan="2">185</td></tr>
<tr><td>Settled amount (Rs)</td><td colspan="2">336,000</td></tr>
</table>

Source: BASIX (2005).

instead like to sell insurance cover to state governments or financial institutions on a wholesale basis.

In 2005, the public sector AIC also launched an indexed insurance scheme for deficit rainfall across 10 states. AIC was able to draw on its established network to sell insurance to more than 125,000 farmers growing crops over 98,000 ha, covering a risk of approx Rs560 million, earning a premium of Rs32 million. Claims were processed in a month from the close of the indemnity period. Compensation of Rs1.2 million was paid to nearly 300 farmers at two stations in Uttar Pradesh.

4.2. Improving access to innovative risk transfer products for the poor

At present, the key challenge faced by the Indian insurance industry is to improve penetration of existing insurance products and to develop innovative delivery mechanisms with a view to improving access of the most vulnerable communities to insurance. This goal can be accomplished through developing better product distribution channels and the design of more cost-effective, and hence more affordable, insurance products. Each of these options is reviewed below.

4.3. Distribution channels

Until recently, the only mode of distribution of insurance products was through the insurance companies' sales force or tied agents. Today, alternative channels such as bancassurance, brokers, corporate agents and direct marketing through the Internet and sales through e-Choupals, MFIs are slowly becoming more popular. Of these, due to its broader reach, bancassurance has the potential to emerge as a significant distribution mechanism, as in the developed countries. Alternative channels of distribution such as bancassurance, direct marketing, Internet sales and telemarketing reduce the costs and enable insurers to reach a wider customer base, primarily low-income groups.

In securing access to the large rural population of the country, microfinance institutions clearly have an important role to play by identifying end-users, understanding their requirements, and employing appropriate methods to extend the products being designed by insurance companies (Sattaiah and Gunaranjan, 2005).

4.4. Product design

One of the key messages from Novib and MIAN (2005) is 'as microfinance has shown that the poor are bankable, microinsurance is showing that they are insurable as well'. However, the critical issue is whether insurance can be made affordable for the poorest and most vulnerable sections of society. Given that up to 45% of premium income in the formal insurance industry is used for costs other than paying claims (e.g. agent commission, operational costs, policy taxes, etc), the large-scale success of insurance products for the poor ('microinsurance') requires a substantial reduction in costs through thoughtful and innovative measures (see Box 1). Products and procedures must be simplified and customized for a target group with irregular income streams.

Special intermediaries are needed between the formal sector and the target group, and these need to deal with moral hazard, design and price appropriate products, and supplement them with education on risk management at the household and enterprise levels. In addition, education and clear communication of insurance products is essential, and particularly so for new weather-indexed products. The need for simple insurance structures and adequate marketing time is important (CRM, 2005). The BASIX experience shows that targeting the right customer groups early on in the process of product marketing can be crucial for its future commercial viability, as shown by the experience with progressive farmers chosen as entry points for the introduction of new initiatives (Lilleor et al., 2005).

Box 1. Microinsurance and climate risk

Microinsurance involves voluntary and contributory schemes at the community level, and has mostly been used for health and funeral expenses. The challenges that climate change poses for conventional insurance are compounded for microinsurance due to the need to deal with dependent risks while maintaining affordable premiums and low transaction costs.

The specific challenges faced by microinsurance providers for climate risks include:

1. the need for specialised actuarial capacity, given the absence of reliable statistical data
2. the time and effort to make potential clients appreciate the benefits of insurance
3. high distribution costs, given a large number of clients paying small premiums for limited coverage.

MFIs providing microinsurance for weather-related risks run the risk of insolvency as long as they tend to operate in limited areas and cannot diversify their lending risks. But they can use weather-indexed insurance to protect their lending portfolio (e.g. BASIX). They can also potentially make use of alternative risk transfer instruments such as catastrophe bonds or reinsurance, but their high costs and the relatively small size of their portfolios may make them unaffordable without assistance from international donors.

The microinsurance delivery models tested in developing countries include direct service delivery to the poor, agent role of MFIs, direct service delivery by an established insurance company, and self-help, mutual, or cooperative models. By acting as an agent, the MFI avoids taking risks for its loan portfolio, as it writes them on the account of the established insurance firm on whose behalf it is acting. However, the products will have to be customized for the poor, which should be the MFIs' responsibility. Similarly, the direct delivery by the insurance company may affect product design due to poor information about client needs, and also potentially high administration costs. The self-help model may also result in considerable risk accumulations and hence increase costs of risk transfer due to the low spread of risk. In India, the DHAN Foundation helps self-help groups (SHGs) in dealing with insurance companies, while Spandana provides insurance directly to the target groups.

Source: Novib and MIAN (2005).

Case studies (SEWA, BASIX) also show that insurance is more effectively sold to the poor as part of other services (credit, farmers' cooperatives, agricultural input supply), rather than as a stand alone product (see Box 2).

Box 2. Integrating microinsurance with other services: a SEWA (Self Employed Women's Association) case study

SEWA offers a broad range of insurance coverage (life, disability, health and property) for one premium. Although it provides life coverage as an agent, all other covers are provided under a full service model. Started in 1992, the scheme offers covers to SEWA members (female self-employed workers) and their husbands.

The rationale behind SEWA's scheme was the almost total absence of any formal security (health, life, property, pension) provisions for women in the informal economy. Conventional insurance companies believed that retailing to this low-income market would result in extremely high transaction costs. The State provided very limited assistance to these women. SEWA initially chose to work exclusively as an agent for insurance companies who had product knowledge, systems and reserves, but has gradually assumed a larger role in direct insurance.

SEWA has been innovative in developing mechanisms to assist members in saving for their premium payments. Linking insurance to the bank products, SEWA has allowed members to save for premiums through fixed deposits into regular savings accounts and also helped it to lock in the membership. The fact that members do not have to make formal insurance premium payments each year is a big incentive to renew insurance covers. However, one of the most important lessons learnt is that clients and staff must know very clearly the details of their coverage. They need to understand not just what is covered, but how to claim, time limits for claims, and the specific terms of coverage.

Members' ability to pay premiums remains a key factor in SEWA's pricing strategy and overall design of the service package. The annual premium is Rs72.5 (US$1.65) for members, and an additional optional Rs22.5 (US$0.51) and Rs20 (US$0.45) for their husbands. Unlike many start-up insurers in the microfinance arena, SEWA maintains significant reserves (provided by GTZ and carrying a principal-plus-inflation maintenance requirement). These reserves provide an important source of revenue through interest earnings that subsidize the programme and its premiums.

The reserves have grown by an average 9% per year, while inflation averaged just under 8% for that period. SEWA has maintained a balance between premiums and levels of coverage by integrating its insurance programme into an array of supporting services. The Bank and housing units support the property insurance. As a stand-alone product, the SEWA insurance is too limited to make a significant impact. However, as a component of an integrated system within the broader spectrum of financial services, SEWA has been able to improve the overall effectiveness of care for its members.

Source: McCord et al. (2001).

5. Concluding thoughts

There is much scope for thinking ahead about the challenges that climate change presents for the Indian insurance sector. There is much to be learnt from the experiences of other countries; for instance, a state-mandated catastrophe insurance pool, like that implemented in Turkey, can add liquidity in the aftermath of a disaster, while also providing the right incentives for mitigation. However, it should also be recognized that insurance can only be part of the solution, and must be combined with genuine preparedness and adaptation activities. The Indian insurance industry should help its customers identify how risks can be managed. It should start planning for a situation where insurance claims may increase with climate change and, at the same time, it should proactively gear up for the potential business opportunities by investing in the development of underwriting and catastrophe risk management skills, designing innovative products and improving access to insurance for under-served customers.

The private insurance industry, however, is limited in its ability to retain highly correlated disaster risk. Some form of government involvement is needed in order to keep the cost of risk financing manageable. Furthermore, only the government can discourage unsustainable economic activities in disaster-prone areas; review coastal land zoning regulation in the light of sea-level rise; and introduce appropriate land-use planning requirements, forest management and mangrove preservation practices. Only the government can mandate or create incentives for purchasing insurance by businesses and (better-off) households by tying disaster insurance to mortgage financing, or to land tax or land registration systems. Only the government can foster efficient government and private insurance risk-sharing through national disaster risk pooling.

In addition, we must point out the importance of prudent insurance regulations for adequate risk management on the part of insurers, and the development of catastrophe risk insurance products. In this context, allowing rate increases and changes in terms and conditions to private operators should be one of the first steps of a prudent insurance regulator to protect the market from the impacts of climate change. One of the regulatory actions should also be to streamline claims-handling procedures in the event of natural disasters as well as raising the level of catastrophe risk underwriting and risk management skills in the insurance industry.

Acknowledgements

This work has been made possible by financial support from the Department of Environment, Food and Rural Affairs (DEFRA), UK. It has benefited from the insightful comments provided by Dr Andrew Dlugolecki, Visiting Fellow, Climatic Research Unit, University of East Anglia. Valuable inputs have also been received from Ms Preety Bhandari, Director, Policy Analysis Division, TERI, and Dr Eugene N. Gurenko, Lead Insurance Specialist of the World Bank. The authors are also grateful to the Infrastructure Regulatory and Development Authority of India for hosting a brainstorming session with several Indian insurance companies in Mumbai in May 2005.

Notes

1 These projections employ the second-generation Hadley Centre Regional Model (Had RM2) and the IS92a future scenarios of increased greenhouse gas concentrations.
2 The insured property losses after the earthquake amounted to a mere 2% of the total estimated property losses (Gurenko, 2004).
3 This phenomenon has also been observed in the USA (Lave and Lave, 1991).
4 Conroy et al. (2001) talk about the increasing importance of remittances from children who have been sent out for jobs to places (often urban) distant from their parents' village. Deshingkar and Start (2003) also show that sending out one or more persons to work in a distant location for part of the year has become a livelihood strategy for many rural households. Rosenzweig and Stark (1989) specifically highlight the role of marrying daughters into families in distant places in order to mitigate covariate risk and to ensure transfers in times of need.

References

Banerjee, A., 2001. Rush for insurance cover. Tribune News Service, Chandigarh edition, 5 February 2001.
BASIX, 2005. Website [available at http://www.basixindia.com/insurance1.asp].
Brooks, N., Adger, W. N., 2003. Country-level risk measures of climate-related natural disasters and implications for adaptation to climate change. Tyndall Centre Working Paper No. 26. Norwich: Tyndall Centre for Climate Change Research.
Conroy, C., Iyengar, S., Lobo, V., Rao, G.B., 2001. Household Livelihood and Coping Strategies in Semi-arid India: Adapting to Long-term Changes. Research Project Report R7558/04. Society for Promotion of Wastelands Development, New Delhi [available at http://www.nrsp.org.uk/database/output_view.asp?outputID=740].

CRM, 2005. Lessons learnt: changing weather patterns. Website of the International Task Force on Commodity Risk Management in Developing Countries [available at http://www.itf-commrisk.org].
Deshingkar, P., Start, D., 2003. Seasonal Migration for Livelihoods in India: Coping, Accumulation and Exclusion. Working Paper No 220. Overseas Development Institute, London.
Government of India, 2004. India's initial national communication to the United Nations Framework Convention on Climate Change. Ministry of Environment and Forests, Government of India, New Delhi [available at http://unfccc.int/resource/docs/natc/indnc1.pdf].
Government of Rajasthan, 2005. Rajasthan State Agricultural Marketing Board website. Government of Rajasthan, Jaipur [available at http://www.rajamb.com/WEATHER_ORANGE.htm].
Gurenko, E., 2004. Catastrophe Risk and Reinsurance: A Country Risk Management Perspective (Ed.). Risk Books, London.
Gurenko, E., Lester, R., 2003. Financing Rapid Onset Natural Disaster Losses in India: A Risk Management Approach. Report No. 26844-IN. The World Bank, Washington, DC [available at http://www-wds.worldbank.org/servlet/WDSContentServer/WDSP/IB/2003/11/10/000090341_20031110134554/Rendered/PDF/268440IN.pdf].
Hess, U., 2003a. Innovative Financial Services for Rural India: Monsoon-Indexed Lending and Insurance for Smallholders. Agriculture and Rural Development Working Paper 9. The World Bank, Washington, DC.
Hess, U., 2003b. Discussant Reaction Paper on Skees, J., 2003. Risk management challenges in rural financial markets: blending risk management innovations with rural finance. Paper presented at Paving the Way Forward for Rural Finance: An International Conference on Best Practices held 2–4 June, Washington, DC [available at http://www.basis.wisc.edu/live/rfc/theme_risk_r3.pdf].
Hess, U., Richter, K., Stoppa, A., 2002. Weather risk management for agriculture and agri-business in developing countries. In: R.S. Dischel (Ed.), Climate Risk and the Weather Market: Financial Risk Management with Weather Hedges. Risk Books, London.
Infrastructure Regulatory and Development Authority Act, 1999. IRDA 2003-04 Annual Report [accessed from www.irdaindia.org/anualreport.html].
IRDA Subregulation 3, 2002. Obligations of Insurers to Rural or Social Sectors, 2002 [accessed from www.irdaindia.org/regulations/gaz-finalregulation ruralsector Obligations.rtf].
Kalavakonda, V., Mahul, O., 2005. Crop Insurance in Karnataka. World Bank Policy Research Working Paper 3654. The World Bank, Washington, DC.
Lave, T.R., Lave, L.B., 1991. Public perception of the risks of floods: implications for communication. Risk Analysis 11(2), 255–267.
Lilleor, H.B., Gine, X., Townsend, R.M., Vickery, J., 2005. Weather insurance in semi-arid India. Paper presented at Securing Development in an Unstable World: Annual Bank Conference on Development Economics 2005, held 23–24 May, Amsterdam [available at http://siteresources.worldbank.org/INTAMSTERDAM/Resources/GinePaper.pdf].
McCord, M.J., Isern, J., Hashemi, S., 2001. Microinsurance: a case study of an example of the full service model of microinsurance provision – Self Employed Women's Association (SEWA). MicroSave, Nairobi, Kenya.
MoHA, 2005. Building a New Techno-legal Regime for Safer India. Ministry of Home Affairs, Government of India, New Delhi.
Munich Re, 2006. Topics Geo: Significant Natural Catastrophes in 2005 [available from http://www.munichre.com].
NCDM and NDMD, 1999. Culture of Prevention: Natural Disaster Management – India. National Centre for Disaster Management, Indian Institute of Public Administration and National Disaster Management Division, Department of Agriculture and Cooperation, Ministry of Agriculture, New Delhi, India.
Novib and MIAN, 2005. Microinsurance: An Overview of Client, Provider, and Support Perspectives. Novib Working Paper No. 1. Novib, Oxfam Netherlands, The Hague [available at www.mian.nl].
OFDA/CRED, 2003. EM-DAT: The OFDA/CRED international disaster database. Universite Catholique de Louvain, Brussels, Belgium [available at http://www.cred.be/emdat].
Rao, C.S., 2004. Indian insurance industry: new avenues for growth. Address given at FICCI, Delhi, 18–19 October 2004.
Rosenzweig, M.R., Stark, O., 1989. Consumption smoothing, migration, and marriage: evidence from rural India. Journal of Political Economy 97(4), 905–926.
Sattaiah, D., Gunaranjan, P.S., 2005. Paper presented at Conference on New Frontiers in Indian Insurance: Rural Insurance, 31 January–1 February 2004, Mumbai. BASIX, Hyderabad [available at http://www.basixindia.com/paper.asp].
Sinha, T., 2005. An Analysis of the Evolution of Insurance in India. CRIS Discussion Papers Series 2005 III. Centre for Risk and Insurance Studies, University of Nottingham, Nottingham, UK [available at http://www.nottingham.ac.uk/business/cris/papers/2005-3.pdf].

Can insurance deal with negative effects arising from climate policy measures?

Axel Michaelowa*

Programme of International Climate Policy, Political Economy of Developing Countries, Institute for Political Science, University of Zurich, Switzerland

Abstract

Articles 4.8 and 4.9 of the United Nations Framework Convention on Climate Change (UNFCCC) and Article 3.14 of the Kyoto Protocol seek to limit adverse effects on developing countries due to the implementation of response measures to climate change, i.e. mitigation and adaptation. In the short term, the availability of renewable energy technology can be affected by mitigation measures. Carbon storage projects can enhance timber supply and reduce revenues from timber sales of other countries. Large-scale technical adaptation programmes can increase prices for commodities used as inputs. Societal adaptation programmes can have negative impacts on neighbouring countries by reducing natural resource availability. However, climate policy measures will not only generate losses but also benefits, often in the same countries that experience losses. Notably, emission reduction measures could reduce import bills (and volatility) for developing countries that import fossil fuels. Losses are unlikely to be insurable due to the characteristics of events with regard to timing, predictability and inseparability of causes. Emitters could be made liable for their emissions, with liability covering direct and indirect effects of the emissions. Market-based financial derivatives allow hedging against fuel price losses, albeit only in the short term. In the long term, the best approach to prevent losses and even reap benefits is an economic diversification. Here the CDM can be used as leverage to mobilize funds, as CER revenues are perfectly negatively correlated with losses from reduced revenues of carbon-rich fuels.

Keywords: Climate policy; Losses; OPEC; Insurance; Diversification; Liability

1. Introduction

Insurance against the negative impacts of climate change is a topic that has been discussed intensively in recent years (see, e.g., Vellinga and Mills, 2001) and which is discussed in several articles in this Special Issue. A much less intensively discussed issue is the question whether insurance can play a role in cushioning possible negative impacts of policies mitigating greenhouse gases and adapting to climate change.

Mitigation of CO_2 emissions is linked to the reduction of fossil fuel use, unless sequestration options are used. The reduction of fossil fuel use has negative impacts on fossil fuel exporters. Therefore, for a long time OPEC countries have argued in the international climate negotiations that they should get a

* Corresponding author. Tel.: +41-43-355-0073
E-mail address: axel.michaelowa@pw.uzh.ch

compensation for reduced export revenues. However, not only emission reduction, but also sequestration and adaptation to climate change, will have secondary effects that will be negative for some stakeholders.

The necessity to address negative impacts of the implementation of mitigation and adaptation policies ('response measures') is specified in Articles 4.8 and 4.9 of the UNFCCC and Article 3.14 of its Kyoto Protocol (for a detailed account of the negotiations on these issues see Barnett and Dessai, 2002). Article 4.8 addresses both climate change impacts and the impacts of climate policies. While listing nine categories of countries that would be particularly subject to impacts of climate change, it specifies two sub-categories of countries subject to impacts of response measures:

- Economies highly dependent on income generated from production, processing and export of fossil fuels
- Economies highly dependent on consumption of fossil fuels and associated energy-intensive products.

Article 4.9 explicitly refers to the need of technology transfer to least-developed countries. Article 3.14 of the Kyoto Protocol echoes the Convention provisions by requiring Annex I Parties to strive to implement their emission targets in such a way as to minimize adverse social, environmental and economic impacts on developing countries.

This article begins with a definition of possible losses from the implementation of climate policy. This is a prerequisite in order to understand the role that insurance or other loss-reducing measures can play. Subsequently, the role of insurance and other policy and market instruments in reducing losses from mitigation and adaptation activities is discussed.

2. Typology of potential negative impacts of climate policy measures on developing countries and quantitative estimates

There is a wide range of climate policy measures with an equally wide range of potentially adverse as well as potentially positive effects on developing countries. The critical issue is that the effects from the impact of the implementation of response measures, for example on terms of trade, international capital flows and development efforts, are unequally distributed and difficult, if not impossible, to quantify. I have used simple, illustrative examples (in boxes) to explain the different negative effects. One might argue that countries benefiting from positive effects would have to deduct them from the gross negative effects before calculating the balance of negative effects.

2.1. Greenhouse gas emission reduction and sequestration

Under the Kyoto Protocol, industrialized countries (Annex I countries) have quantified emission limitation and reduction obligations and thus will embark on mitigation action. The necessary scale of mitigation depends on a number of factors such as economic growth, technological development, supply of different fuels and availability of mitigation options worldwide (e.g. through the Clean Development Mechanism, CDM). After the USA's decision not to ratify the Kyoto Protocol and the Marrakech Accords, some model runs on the global greenhouse gas market have been done (Eyckmans et al., 2001; Blanchard et al., 2002; den Elzen and Both, 2002; Jotzo and Michaelowa, 2002). Most of these models conclude that the world market price of CO_2 would be zero if the countries in transition sell all their surplus permits. If these suppliers restrain sales, the world market price would reach about €2–5 per tonne CO_2

equivalent. However, current models are not able to model climate policy impacts adequately. Results vary depending on the model used and on input data or assumptions. Existing models produce a wide diversity of short-term impacts, although if a full portfolio of mitigation options is used, all models show that potential adverse impacts would be reduced. In the longer term (post-2020), with more aggressive reductions, impacts may be greater – although this depends on policy choices.

So far, developing countries do not have any obligations to limit or reduce emissions. Thus their activities are likely to focus on voluntary mitigation measures with economic benefits ('no regret' measures) such as the abolition of fossil fuel subsidies.

2.1.1. Reduction of demand for carbon-rich fuels

Any mitigation action in the energy sector will lead to a reduced demand for fossil fuels which falls particularly heavily on fuels with a high carbon content (Bartsch and Müller, 2000). This in turn will reduce world market prices for these fuels. The fall in price and export volume will reduce revenues of *fossil-fuel-exporting* countries. Countries *importing* fossil fuels will unambiguously profit from the lower prices.

Due to the wide range of parameters that influence energy markets, it is impossible to unambiguously separate the price and quantity effect caused by mitigation.

Box 1. Loss through reduction of demand for carbon-rich fuels

> Due to the conversion of the electricity generation system from coal to wind in Annex B country Aeolia, the coal exports from developing country Carbostan to Aeolia drop from 10 million t per annum to zero. Likewise, the coal market price falls from €20 to €10 per t. Due to long-term export contracts with the other importers, overall coal exports from Carbostan only fall from 50 million to 40 million t but prices have to be adjusted. Carbostan claims a loss of €600 million (€1,000 million previous coal export revenues compared to €400 million after Aeolia's action) due to mitigation.

2.1.2. Increased demand for renewable energy and energy efficiency technologies

Mitigation action will mean an increased investment in renewable energy/energy efficiency technology. A lower availability and higher price of such technologies for developing countries is possible, if supply cannot cope with demand, in the short term. For example, PV module prices have not fallen in the last few years in the highly subsidized markets of Germany and Japan. German wind turbine producers shunned export markets for a long time due to the high demand in their home market. However, these subsidies surpass projected market prices by several orders of magnitude and it is clear that investment at rates derived from market prices will not lead to diversion of renewable technology exports. In the long run, positive impacts from increased renewables investment will result. Long-term availability of renewable energy efficiency technologies will be enhanced due to economies of scale that lead to lower prices.

Box 2. Loss through increased demand for mitigation technologies

> Due to a new 10,000 MW wind programme in Annex B country Aeolia, all Aeolian wind turbine manufacturers operate at full capacity. Thus the developing country Tempesto cannot place an order for 100-MW wind turbines with a producer in Aeolia. It thus has to switch to a turbine producer in Breezia which charges a price of €1,200 per kW installed instead of the Aeolian producers' list price of €1,000 per kW. Tempesto claims a loss of €20 million (€200,000/MW times 100 MW) due to Aeolia's mitigation.

2.1.3. Timber market effects due to carbon sequestration in terrestrial vegetation

Carbon sequestration through afforestation and deforestation is allowed under the CDM. Increased forestry activities will lead to a decrease in timber prices in the future due to the enhanced supply. This will be negative for timber exporters and positive for timber importers.

Box 3. Loss through timber market effects

> In 2005, the developing country Arboria approved a CDM afforestation project on 100,000 ha whose first harvest occurs in 2020. In 2020, the country Verdura logs 10,000 ha and harvests 1 million t of timber. Due to Arboria's timber supply, timber prices fall from €50/t to €45/t. Verdura claims a loss of €5 million (€5/t times 1 million t) due to Arboria's sequestration.

2.2. Adaptation to climate change

Many developing countries have an interest in adaptation if they can reduce the net costs of climate change impacts. As adaptation can be anything that enhances the resilience of a society (and is thus correlated with the degree of development of a society in general), this article will concentrate on the main options. It will not focus on the costs of adaptation but on the impacts of adaptation measures on other countries.

2.2.1. Technical adaptation to meteorological extremes

Adaptation to sea-level rise and enhanced flooding will generate expenses for coastal and riparian protection – such as dykes, shelters, pumps and sluice-gates. These expenses will stimulate the construction industry and raise prices for corresponding inputs. As construction is a localized industry, there will not be many direct adverse impacts for other countries.

Box 4. Loss due to technical adaptation

> In 2010, the developing country Neptunia builds a seawall on 500 km of coastline. Due to the high demand for building material, export prices for 1 million t of cement to neighbouring Montania increase from €55/t to €65/t. Montania claims a loss of €10 million (€10/t times 1 million t) due to Neptunia's adaptation.

2.2.2. Societal adaptation

Adaptation not only entails technological fixes but also increases in flexibility in order to lower the susceptibility to climatic extremes. This entails expenses for early warning systems and for institutions such as agricultural and forestry extension services that give recommendations on good agricultural practices and train farmers/foresters. Management of certain natural resources such as irrigation and hydropower systems or skiing resorts will have to be changed. Direct adverse effects of such activities are unlikely; in some circumstances indirect effects on resource availability could result (see Box 5). Normally, societal adaptation will have benefits concerning other economic activities, and thus ancillary benefits. For example, an improved agricultural extension service will also be useful to deal with a new pest infestation that has nothing to do with climate change.

Box 5. Loss due to societal adaptation

> In 2010, the developing country Fluvia introduces a new operation plan for its irrigation system to be able to withstand more severe droughts due to projected climate change. Due to the much lower cost of irrigation, farmers expand irrigation and consequently the amount of water discharged to neighbouring Desertum declines by 10%. Desertum argues that it has to reduce its irrigated area by 100,000 ha and claims a loss of €10 million (€100/ha times 100,000 ha) due to Fluvia's adaptation.

A general problem to be addressed in all discussion of reducing losses from mitigation and adaptation activities is the problem of separating the effects of climate policy measures from 'noise' generated by socio-economic development. In the short run, the impact of market effects on prices and revenues from fossil fuel exports is likely to dominate eventual price and revenue effects from implementation of climate policy. The drastic oil price increase of the last years is much higher than any price reductions envisaged due to greenhouse gas reduction measures. It could even be argued that costs incurred to achieve greenhouse gas reductions could 'shave' future sudden price increases (with the ensuing costly emergency oil savings measures in importing countries) and thus be seen as an 'insurance premium' to attenuate upward price risks.[1]

3. Principal options for alleviating losses

Losses can be addressed in different ways. Depending on the parameters influencing size, distribution and timing of losses, financial instruments or insurance-related instruments can be applied.

3.1. Provision of insurance

Insurance is possible if risks (i.e. probabilities of a damaging event) can be assessed in a systematic way. Moreover, risks must be spread across a large set of entities and not occur at the same time. In the case of losses from the implementation of climate policy measures, the characteristics of the event triggering insurance payments differ from normal insurable events. In the latter case, past experience has generated a probability distribution of a loss-triggering event. If events happen that go beyond this experience, premiums are adjusted and insurance coverage reduced, as was the case on hurricane-stricken islands in the Caribbean.

A necessary condition for insurance is a set of suitable definitions of loss-triggering events (see Box 6).

Box 6. Possible definitions of loss-generating events

> An insurance payment could become due
> - when Annex B country Aeolia's wind power capacity has reached 20 GW
> - when coal world market prices fall below €20/t
> - when annual coal export revenues of developing country Carbostan fall below €300 million
> - when prices for wind turbines rise above €1,200/kW
> - when developing country Neptunia completes 500 km of sea-wall.

However, the following event characteristics can be found that have a negative impact on insurability:

- Most of the possible impacts from the implementation of climate policy are fairly evenly spread but could occur simultaneously. Thus inter-temporal diversification of risk is not possible.
- Policy implementation is often predictable regarding its time. In this case, insurers would simply charge the premium commensurate with this risk on the year when it materializes. Insuring against discrete policy actions such as the building of a sea-wall would require premiums that are prohibitively high.
- Implementation of emission reduction measures invariably reduces prices of carbon-rich fuels but the effect will be masked by a multitude of other influences. The insurance premiums would have to cover all those influences and would thus become very high.
- Only rarely will there be one distinct event. Normally many small activities will have a gradual influence over time.
- Adverse effects from adaptation action can be specified more clearly, as the adaptation action has a clear starting and end point.

Overall, these characteristics mean that losses from climate policy measures are unlikely to be insurable.

Even uninsurable losses are often spread through public funds financed essentially out of taxpayers' contributions (Vellinga and Mills, 2001, p. 436). However, for climate policy damages in other countries, taxpayers' willingness to pay will be extremely low. If one uses the development assistance budgets on the industrialized countries as an indicator for willingness to pay, the consistent decline in these budgets over the past decade is a clear sign.

3.2. Allocation of liability

Linnerooth-Bayer et al. (2003) summarize the currently operational insurance mechanisms for liability and damage. Losses from climate policy actions could be covered if greenhouse gas emitters were liable for adverse effects from their emissions (Germanwatch, 2003). While it is not possible to prove a direct link between the emission of a certain amount of greenhouse gases and a specific loss event, one can argue that due to the global mixing of greenhouse gases, the allocation of liability is proportional to the emissions share. Tol and Verheyen (2004) argue that a country has to pay compensation if an 'activity under its control does damage to another country, and if this is done on purpose or due to carelessness'. A country would thus be liable for emitters on its territory. Making such emitters liable is a process that can take decades, as the cases of asbestos and tobacco production have shown, so immediate results are unlikely. However, insurers have been reported to begin to exclude risks from eventual future emissions liabilities in company directors-and-officers liability contracts (Ball, 2003). Liability could be defined as covering the *direct effects* of the emissions (climate change impacts) as well as costs of *measures aimed at preventing the direct effects* (mitigation and adaptation). The latter costs would include losses incurred to third parties from the implementation of mitigation and adaptation (see Figure 1). Obviously under the 'polluter pays' and the precautionary principle, the liability for the direct effects takes precedence and should be prioritized, but a consistent treatment of all adverse effects of climate change would facilitate efforts to achieve a long-term, universal climate policy regime.

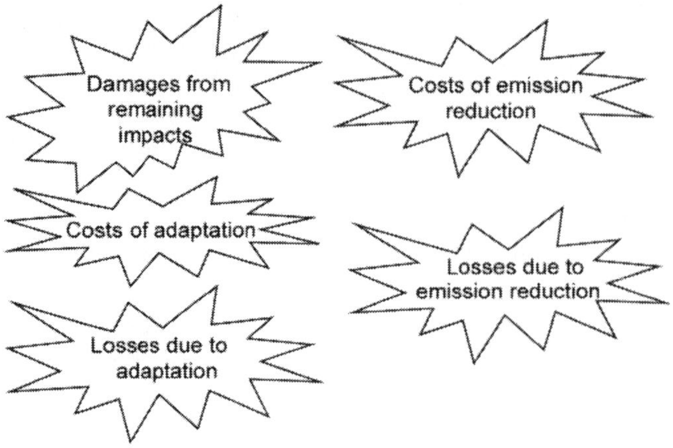

Figure 1. Different levels of liability.

The main problems are the enforcement of liability (country level, emitter level, consumer level) and its quantification. As third party losses from the implementation of mitigation and adaptation would accrue quickly, the temporal problem related to climate change impacts would not exist here. Liable actors are likely to insure themselves. However, current environmental insurance against third party liability is only done after a very careful risk analysis and against sudden and accidental occurrences only (UNEP Finance Initiatives, 2002, p. 37).

3.3. Financial derivatives

Capital markets have a depth far exceeding the resources available to insurance and reinsurance providers. There is a broad range of financial derivatives covering different lifetimes to guarantee revenue from commodity exports. Companies regularly hedge their sales. Standardized futures and options are traded on organized commodity exchanges. For metals, petroleum products and certain agricultural products (e.g. coffee, cocoa, soybean and soybean products, wheat, maize) there are relatively liquid markets, and for several commodities traded on international exchanges, the volume of contracts traded is several times the volume of physical production (International Task Force on Commodity Risk Management, 2003b). However, they are only available for relatively short periods and for homogeneous commodities. Thus their applicability to buffering losses from climate policy measures is limited. Larson et al. (1998, pp. 22ff) describe examples of how oil-exporting countries use these instruments to hedge against price decreases.

Over-the-counter (OTC) instruments are bilaterally agreed risk management instruments that are traded outside of the organized exchanges. OTC options and long-dated swaps[2] can cover multi-year periods. They are common in the oil market, covering periods of 5–7 years (UNCTAD, 1998a, p. 38). Thus they can form models for the reduction of losses from climate policy measures. Table 1 summarizes the different instruments.

The longer the lifetimes, the less standardized instruments become and the higher transaction costs are. Of course, market expectations play an important role in pricing of all instruments. If the market expects a quick implementation of climate policy and a high probability of fossil fuel price

Table 1. Financial derivatives for price hedging

Instrument	Characteristics	Lifetime	Applicability in climate policy context
Forward	Price fixed at future date, traded OTC. Physical delivery expected.	<1 year	None
Future	Price fixed at future date, traded on an exchange. Less risky than forward due to margin payments.[a]	<3 years	Limited
Option	Price fixed at future date, traded OTC or on an exchange. Premium paid at beginning.	<3 years	Limited
Swap	Exchange of specified cash flows at specific intervals (series of forwards). Tailor-made contracts	<25 years, normally <7 years	High: can cover multiple commitment periods
Commodity bond	Repayment linked to price. Tailor-made contracts	<30 years	High: can cover multiple commitment periods

[a] A margin payment is a payment required if the futures price moves across a specific threshold.
Source: for a detailed description of types, see UNCTAD (1998a).

decrease, premiums for options/interest rates for commodity loans and bonds will rise. The current high price level of fossil fuels due to the political insecurity in the Middle East may be a window of opportunity to negotiate contracts at attractive conditions with a duration that is as long as possible.

Market-based financial derivatives can be dangerous if improperly applied. Especially for complex deals, there is often an information imbalance: the fair costs of structuring the financing, or of the risk management instrument, are only known to the provider. Furthermore, if an inappropriate strategy is chosen (i.e. a strategy that is more complex than the entity can handle), the risks of using derivatives markets can be large. Using futures and over-the-counter risk management markets in a more active manner requires a good and relatively sophisticated control environment (UNCTAD, 1998b). Country officials of fossil-fuel-exporting countries are usually well-trained in the application of financial hedging instruments. For less-developed countries, training courses in using financial derivatives could be an option, although rapid changes in the design of derivatives tend to lead to a quick obsolescence of acquired competencies, unless used regularly.

3.4. Savings and stabilization funds

Revenues from the export of commodities can be collected in times of high prices and be distributed in periods of low prices or declining overall production. Over the past half century, the international community has come up with stabilization or compensatory mechanisms to help developing countries alleviate the negative impact of commodity price fluctuations. These schemes have typically taken the form of institutional arrangements for price stabilization programmes, including physical buffer stock schemes, stabilization funds or variable tariff schemes, and marketing boards. But many of these programmes which attempted to separate domestic commodity prices from international prices over time often proved financially unsustainable. Many of the schemes failed because they were based on administratively set benchmarks which required large resource transfers in years of low prices. With limited borrowing capacity and generally unhedged exposure to price risks, the stabilization programmes were difficult to maintain when payments were required over consecutive years. The stabilization components of the international commodity agreements also proved unsustainable and are no longer in force (International Task Force on Commodity Risk Management, 2003a). More success has been achieved by accumulating savings whose returns were then used to substitute for export revenues.

Several governments use oil funds; some for savings, others for stabilization purposes (see Table 2). Looking at the oil market in Venezuela, Claessens and Varangis (1994) found that for a stabilization fund to be effective, several preconditions must be met. Most notably, oil prices should not have a systematic trend, financial markets should be incomplete, and there are large adjustment costs. Stabilization funds can only function if managed in a professional way; rules should not be changed often (Fasano, 2000). Revenues of such a fund could be used for diversification. However, policymakers always have an incentive to spend money in a fund accumulated by their predecessors.

The Norwegian oil fund is designed to fund pensions in the face of declining oil revenues over decades, and thus has an aim similar to multi-decadal revenue stabilization in the face of climate policy. It spreads its investments in equities and bonds on a global scale (Government of Norway, 2003). A fund will only be sustainable if it is managed in a transparent manner. Governments that use funds to accumulate fossil fuel export revenue could disseminate their knowledge about an effective management of such funds.

Table 2. Oil funds

Name	Country	Start date	Value 2002 (€ billion)
Government Petroleum Fund	Norway	1990	80
Fund for Future Generations	Kuwait	1960	Peak 100[a] 40
Alaska Permanent Reserve Fund	USA	1976	20
Alberta Heritage Savings Trust Fund	Canada	1976	8
Investment Fund for Macroeconomic Stabilization	Venezuela	1999	3.5
State General Reserve Fund	Oman	1980	2
Foreign Exchange Reserve Account	Iran	1999	1
National Fund	Kazakhstan	2000	1
State Oil Fund	Azerbaijan	1999	0.5

[a] Before the Iraqi invasion in 1990 (KUNA, 2003).
Source: Melby (2002).

3.5. Diversification

The best long-term risk management policy for countries exporting fossil fuels is to diversify away from commodities that run price and demand risks. Funds for diversification could be raised through taxes on the production of fossil fuels, whose revenues would be earmarked for diversification projects.

Many fossil-fuel-exporting countries have a good renewable energy resource base, especially concerning solar, but also wind, energy. Diversification in this direction would reduce several risks at the same time. Moreover, the CDM can be harnessed to provide funding and technology for renewable energy deployment. As the CDM incentive will only be getting stronger with the rise of carbon price in the global market, it will also have a close negative correlation with the amount of losses due to emission reduction measures. Nevertheless, fossil fuel exporters so far have neither taken up the opportunities of the pilot phase of Activities Implemented Jointly, nor have they made visible efforts in the CDM area. Only very recently, some fossil fuel exporters have set up their national approval authorities and embarked on awareness-raising among their private and public sectors.

Another direction for diversification is the production of energy-intensive commodities. As energy prices rise in Annex B countries, they will reduce domestic production of such commodities and increase imports. However, if the climate policy regime becomes universal, countries that have specialized in energy-intensive production would become subject to emission reduction requirements.

Investment in new technologies to reduce the costs of geological carbon sequestration may lead to a situation where the carbon penalty of fossil fuels loses its relevance if sequestration becomes cost-competitive with emission reduction through renewable energy or energy efficiency.

Hence, diversification is the only viable strategy for the very long term, as hedging instruments at a certain point of time will not be available for more than two decades into the future. The successful strategy of Dubai in setting up free-trade zones and diversifying into services could serve as a blueprint.

4. Conclusions

The role of formal insurance in reducing losses of fossil-fuel-exporting countries from global climate change adaptation measures seems to be extremely limited due to the characteristics of events with regard to timing, predictability and separability of causes. In the short term (up to 3 years), financial derivatives can be used to guarantee prices of carbon-rich export commodities unless the market has already depressed the price. In the medium term, up to a decade, commodity bonds may fulfil the same function but are difficult to negotiate.

The only viable long-term strategy in this regard is economic diversification, where harnessing of CDM funds can provide a leverage that counteracts negative price effects on carbon-rich commodities. The impact of global climate change adaptation measures on fossil exporters, however, can also be somewhat reduced if cost-effective and irreversible storage technologies can be developed in the area of geological carbon sequestration which would allow the continued use of fossil fuels.

Acknowledgements

An earlier version of this article was presented at the UNFCCC Workshop on Insurance-related Actions to Address the Specific Needs and Concerns of Developing Country Parties Arising from the Adverse Effects of Climate Change and from the Impact of the Implementation of Response Measures, Bonn, 14–16 May 2003. I thank three anonymous referees and the editor of this Special Issue for valuable comments.

Notes

1 I thank an anonymous referee for making that point.
2 A swap is an instrument where you agree to exchange two different commodities at a point of time in the future.

References

Ball, J., 2003. Insurers weigh moves to cut liability for global warming. Wall Street Journal Europe, 7 May 2003.

Barnett, J., Dessai, S., 2002. Articles 4.8/4.9 of the UNFCCC: The adverse effects and the impacts of response measures. Climate Policy 2(3), 231–239.

Bartsch, U., Müller, B., 2000. Fossil Fuels in a Changing Climate. Oxford University Press, Oxford, UK.

Blanchard, O., Criqui, P., Kitous, A., 2002. After The Hague, Bonn and Marrakech: the future international market for emissions permits and the issue of hot air. Cahier de Recherche 27bis, IEPE, Grenoble.

Claessens, S., Varangis, P., 1994. Oil Price Instability, Hedging, and an Oil Stabilization Fund: The Case of Venezuela. Policy Research Working Paper WPS 1290. The World Bank, Washington, DC.

den Elzen, M.G.J., Both, S., 2002. Modelling Emissions Trading and Abatement Costs in Fair 1.1 – Case Study: The Kyoto Protocol under the Bonn–Marrakesh Agreement. RIVM Report 728001021/2002. Bilthoven, The Netherlands.

Eyckmans, J., Van Regemorter, D., van Steenberghe, V., 2001. Is Kyoto Fatally Flawed? An Analysis with MacGEM. Working Paper 2001-18. Nota de Lavoro 43.2002. Centre for Economic Studies, Catholic University of Leuven, Leuven, Belgium.

Fasano, U., 2000. Review of the Experience with Oil Stabilization and Savings Fund in Selected Countries. IMF Working Paper 00/112. IMF, Washington, DC.

Germanwatch, 2003. The Climate Responsibility Campaign [available at http://www.germanwatch.org/klak/intro-e.htm, accessed May 8, 2003].

Government of Norway, 2003. Government of Norway Petroleum Fund [available at http://odin.dep.no/fin/engelsk/p10001617/p10002780/007051-990004/index-dok000-b-n-a.html, accessed 4 May 2003].

International Task Force on Commodity Risk Management, 2003a. Need for price insurance [available at http://www.itf-commrisk.org/itf.asp?page=91, accessed 3 May 2003].

International Task Force on Commodity Risk Management, 2003b. Implementation of price risk management [available at http://www.itf-commrisk.org/itf.asp?page=93, accessed 3 May 2003].

Jotzo, F., Michaelowa, A., 2002. Estimating the CDM market under the Marrakech Accords. Climate Policy 2, 179–196.

KUNA, 2003. Economy and Business [available at http://www.kuna.net.kw/e_c_o_n_o_m_y__a_n_d__b_u_s_i_n_.htm, accessed 4 May 2003].

Larson, D., Varangis, P., Yabuki, N., 1998. Commodity risk management and development. Paper prepared for the Round Table Discussion on New Approaches to Commodity Price Risk Management in Developing Countries, Washington, DC, 28 April 1998. The World Bank, Washington, DC.

Linnerooth-Bayer, J., Mace, M., Verheyen, R., 2003. Insurance-related actions and risk assessment in the context of the UNFCCC. Commissioned by UNFCCC Secretariat, Bonn.

Melby, E.D.K., 2002. A global overview of oil funds. Prepared for the IGAD Symposium. Paper presented at Sudan Peace Talks, Machakos [available at www.csis.org/africa/0208_SudanMelby.pdf, accessed 6 May 2003].

Tol, R., Verheyen, R., 2004. State responsibility and compensation for climate change damages: a legal and economic assessment. Energy Policy 32, 1109–1130.

UNCTAD, 1998a. A Survey of Commodity Risk Management Instruments. UNCTAD/COM/15/Rev.2, Geneva, Switzerland.

UNCTAD, 1998b. Examination of the Effectiveness and Usefulness for Commodity-dependent Countries of New Tools in Commodity Markets: Risk Management and collateralized Finance. TD/B/COM.1/EM.5/2, Geneva, Switzerland.

UNEP Finance Initiatives, 2002. Industry as a Partner for Sustainable Development. Finance and Insurance, Geneva, Switzerland.

Vellinga, P., Mills, E., 2001. Insurance and other financial services. In: IPCC (Ed.), Climate Change 2001: Impacts, Adaptation and Vulnerability. Cambridge University Press, Cambridge, UK, pp. 417–450.

Conclusions and recommendations

Eugene N. Gurenko

World Bank, Washington DC, US

1. Climate change and insurance in developing countries

Climate change is becoming an important factor in the economic and social development of many countries which can no longer be ignored or viewed as incidental. If the previous loss statistics are anything to go by, future economic losses from extreme weather events are only likely to get worse with the acceleration of climate change. Although the adverse impact of climate change may vary from one country to another, it is developing countries that are likely to be affected most due to the lack of fiscal resources and effective national risk adaptation mechanisms, such as insurance.

As well as compensating for the loss and reducing the uncertainty of fiscal outcomes for governments, insurance can be instrumental to adequately assessing, managing and reducing countries' risk exposures to the risk of natural disasters. However, today access to insurance for climate-related risks is extremely limited in developing countries, which leaves their governments, businesses and individuals extremely vulnerable to natural disasters.

2. Insurance-based mechanisms and the global dialogue on adaptation to climate change

In the global dialogue on climate change and disaster risk management, governments, international organizations and other stakeholders should seriously explore the potential of insurance-based mechanisms to spread and reduce losses from extreme weather events. The Munich Climate Insurance Initiative (MCII), comprising international financial organizations, reinsurers, think-tanks and NGOs is the first example of an international initiative which has brought insurance solutions to the forefront in the agenda of the global dialogue on adaptation to climate change.

The aforementioned MCII efforts in promoting global insurance-based climate change adaptation strategies is only the first step forward on the long road ahead. Based on the risk management needs of countries and communities at risk, a wider coalition of interested bodies should be formed to advance the climate insurance agenda more decisively. In particular, the UNFCCC process, including the subsidiary bodies on science, technology and implementation, should make insurance-based climate adaptation mechanisms an integral part of their agenda. Consideration of the scientific and technical aspects of climate insurance should also become a central theme in the new 5-year work on adaptation in the SBSTA. The role of climate-based insurance should also be considered in the negotiations concerning the post-2012 regime under the Kyoto Protocol.

3. Public–private partnerships in climate insurance

Due to the enormous and frequently unpredictable loss accumulation potential from climate-related events, private insurers and reinsurers are constrained in their ability to offer catastrophe risk coverage on weather-related perils world-wide. Yet, the reluctance of the private insurance industry to commit its capital capacity and underwriting expertise to climate-related perils in certain parts of the world can be overcome by forming public–private partnerships. Such an insurance-based approach to financing the risk of climate change can become an attractive alternative to the current highly inefficient *ad hoc* model of international financial assistance to developing countries in the aftermath of natural disasters. In addition, due to their ability to understand and price the risk of climate risk, insurers can help assess and price the risk of extreme weather events for the concerned countries. This information can be then used by governments and international financial organizations to build equitable and effective climate-change adaptation strategies, including global compensation programmes for countries that are likely to be most adversely affected by climate change in the future.

One of the key advantages of public–private partnerships in climate-based insurance is their ability to create strong incentives for proactive risk reduction measures on the part of disaster-prone countries. In the absence of *ex-ante* climate adaptation policies at the country level, which are currently discouraged by the existing system of post-disaster donor aid, the countries' vulnerabilities to weather risks are only likely to increase in the future.

Although insurance-based solutions have clear advantages over non-contractual post-disaster international assistance to developing countries, insurance can be too costly, particularly for countries that are poor and at the same time highly prone to natural disasters. These affordability constraints, however, can be successfully addressed through the creation of insurance-based public–private partnerships that would rely on free risk capital provided by the international donor community, and which may subsidize the development of insurance infrastructure and insurance premiums for the poorer nations.

climate policy

INDEX TO VOLUME 6 (2006)

Title index for volume 6 (2006)

Keyword index for volume 6 (2006)

Author index for volume 6 (2006)

FSC
www.fsc.org
MIX
Papier | Fördert gute Waldnutzung
FSC® C083411

Kolibri 360 GmbH
Ferdinand-Jühlke-Straße 7
99095 Erfurt, Deutschland
produktsicherheit@kolibri360.de